HARPER TORCHBOOKS / The Cloister Library

Tor Andrae	MOHAMMED: *The Man and His Faith* TB/62
Augustine/Przywara	AN AUGUSTINE SYNTHESIS TB/35
Roland H. Bainton	THE TRAVAIL OF RELIGIOUS LIBERTY TB/30
Karl Barth	DOGMATICS IN OUTLINE TB/56
Karl Barth	THE WORD OF GOD AND THE WORD OF MAN TB/13
Nicolas Berdyaev	THE BEGINNING AND THE END TB/14
Nicolas Berdyaev	THE DESTINY OF MAN TB/61
James Henry Breasted	DEVELOPMENT OF RELIGION AND THOUGHT IN ANCIENT EGYPT TB/57
Martin Buber	ECLIPSE OF GOD: *Studies in the Relation Between Religion and Philosophy* TB/12
Martin Buber	MOSES: *The Revelation and the Covenant* TB/27
Jacob Burckhardt	THE CIVILIZATION OF THE RENAISSANCE IN ITALY [Illustrated Edition]: *Vol. I*, TB/40; *Vol. II*, TB/41
Edward Conze	BUDDHISM: *Its Essence and Development* TB/58
F. M. Cornford	FROM RELIGION TO PHILOSOPHY: *A Study in the Origins of Western Speculation* TB/20
G. G. Coulton	MEDIEVAL FAITH AND SYMBOLISM TB/25
G. G. Coulton	THE FATE OF MEDIEVAL ART IN THE RENAISSANCE AND REFORMATION TB/26
H. G. Creel	CONFUCIUS AND THE CHINESE WAY TB/63
Adolf Deissmann	PAUL: *A Study in Social and Religious History* TB/15
C. H. Dodd	THE AUTHORITY OF THE BIBLE TB/43
Johannes Eckhart	MEISTER ECKHART: A Modern Translation TB/8
Mircea Eliade	COSMOS AND HISTORY: *The Myth of the Eternal Return* TB/50
Morton S. Enslin	CHRISTIAN BEGINNINGS TB/5
Morton S. Enslin	THE LITERATURE OF THE CHRISTIAN MOVEMENT TB/6
Austin Farrer, ed.	THE CORE OF THE BIBLE TB/7
Ludwig Feuerbach	THE ESSENCE OF CHRISTIANITY TB/11
Harry Emerson Fosdick	A GUIDE TO UNDERSTANDING THE BIBLE TB/2
Sigmund Freud	ON CREATIVITY AND THE UNCONSCIOUS: *Papers on the Psychology of Art, Literature, Love, Religion* TB/45
Maurice Friedman	MARTIN BUBER: *The Life of Dialogue* TB/64
Octavius Brooks Frothingham	TRANSCENDENTALISM IN NEW ENGLAND: *A History* TB/59
Edward Gibbon	THE END OF THE ROMAN EMPIRE IN THE WEST [J. B. Bury Edition; Illustrated] TB/37
Edward Gibbon	THE TRIUMPH OF CHRISTENDOM IN THE ROMAN EMPIRE [J. B. Bury Edition; Illustrated] TB/46
Charles C. Gillispie	GENESIS AND GEOLOGY TB/51
Maurice Goguel	JESUS AND THE ORIGINS OF CHRISTIANITY I: *Prolegomena to the Life of Jesus* TB/65
Maurice Goguel	JESUS AND THE ORIGINS OF CHRISTIANITY II: *The Life of Jesus* TB/66
Edgar J. Goodspeed	A LIFE OF JESUS TB/1
Herbert J. C. Grierson	CROSS-CURRENTS IN 17TH CENTURY ENGLISH LITERATURE: *The World, the Flesh, the Spirit* TB/47
William Haller	THE RISE OF PURITANISM TB/22
Adolf Harnack	WHAT IS CHRISTIANITY? TB/17
Edwin Hatch	THE INFLUENCE OF GREEK IDEAS ON CHRISTIANITY TB/18
Karl Heim	CHRISTIAN FAITH AND NATURAL SCIENCE TB/16
F. H. Heinemann	EXISTENTIALISM AND THE MODERN PREDICAMENT TB/28
Stanley R. Hopper, ed.	SPIRITUAL PROBLEMS IN CONTEMPORARY LITERATURE TB/21
Johan Huizinga	ERASMUS AND THE AGE OF REFORMATION TB/19
Aldous Huxley	THE DEVILS OF LOUDUN: *A Study in the Psychology of Power Politics and Mystical Religion in the France of Cardinal Richelieu* TB/60
Søren Kierkegaard	EDIFYING DISCOURSES: A Selection TB/32

(Continued on next page)

Søren Kierkegaard	THE JOURNALS OF KIERKEGAARD TB/52
Søren Kierkegaard	PURITY OF HEART TB/4
Alexandre Koyré	FROM THE CLOSED WORLD TO THE INFINITE UNIVERSE TB/31
Emile Mâle	THE GOTHIC IMAGE: *Religious Art in France of the 13th Century* TB/44
H. Richard Niebuhr	CHRIST AND CULTURE TB/3
H. Richard Niebuhr	THE KINGDOM OF GOD IN AMERICA TB/49
H. J. Rose	RELIGION IN GREECE AND ROME TB/55
Josiah Royce	THE RELIGIOUS ASPECT OF PHILOSOPHY: *A Critique of the Bases of Conduct and of Faith* TB/29
Auguste Sabatier	OUTLINES OF A PHILOSOPHY OF RELIGION BASED ON PSYCHOLOGY AND HISTORY TB/23
George Santayana	INTERPRETATIONS OF POETRY AND RELIGION TB/9
George Santayana	WINDS OF DOCTRINE *and* PLATONISM AND THE SPIRITUAL LIFE TB/24
Friedrich Schleiermacher	ON RELIGION: *Speeches to Its Cultured Despisers* TB/36
Henry Osborn Taylor	THE EMERGENCE OF CHRISTIAN CULTURE IN THE WEST: *The Classical Heritage of the Middle Ages* TB/48
Paul Tillich	DYNAMICS OF FAITH TB/42
Edward Burnett Tylor	THE ORIGINS OF CULTURE TB/33
Edward Burnett Tylor	RELIGION IN PRIMITIVE CULTURE TB/34
Evelyn Underhill	WORSHIP TB/10
Johannes Weiss	EARLIEST CHRISTIANITY: *A History of the Period* A.D. *30–150*: Vol. I, TB/53; Vol. II, TB/54
Wilhelm Windelband	A HISTORY OF PHILOSOPHY I: *Greek, Roman, Medieval* TB/38
Wilhelm Windelband	A HISTORY OF PHILOSOPHY II: *Renaissance, Enlightenment, Modern* TB/39

HARPER TORCHBOOKS / The Academy Library

H. J. Blackham	SIX EXISTENTIALIST THINKERS TB/1002
Walter Bromberg	THE MIND OF MAN: *A History of Psychotherapy and Psychoanalysis* TB/1003
G. P. Gooch	ENGLISH DEMOCRATIC IDEAS IN THE SEVENTEENTH CENTURY TB/1006
Francis J. Grund	ARISTOCRACY IN AMERICA TB/1001
Henry James	THE PRINCESS CASAMASSIMA TB/1005
Georges Poulet	STUDIES IN HUMAN TIME TB/1004

HARPER TORCHBOOKS / The Science Library

J. Bronowski	SCIENCE AND HUMAN VALUES TB/505
W. C. Dampier, ed.	READINGS IN THE LITERATURE OF SCIENCE TB/512
Arthur Eddington	SPACE, TIME AND GRAVITATION: *An Outline of the General Relativity Theory* TB/510
H. T. Pledge	SCIENCE SINCE 1500: *A Short History of Mathematics, Physics, Chemistry, and Biology* TB/506
George Sarton	ANCIENT SCIENCE AND MODERN CIVILIZATION TB/501
Paul A. Schilpp, ed.	ALBERT EINSTEIN: Philosopher-Scientist: Vol. I, TB/502; Vol. II, TB/503
Friedrich Waismann	INTRODUCTION TO MATHEMATICAL THINKING: *The Formation of Concepts in Modern Mathematics* TB/511
W. H. Watson	ON UNDERSTANDING PHYSICS: *An Analysis of the Philosophy of Physics* TB/507
G. J. Whitrow	THE STRUCTURE AND EVOLUTION OF THE UNIVERSE: *An Introduction to Cosmology* TB/504
A. Wolf	A HISTORY OF SCIENCE, TECHNOLOGY AND PHILOSOPHY IN THE 16TH AND 17TH CENTURIES: Vol. I, TB/508; Vol. II, TB/509

INTRODUCTION TO MATHEMATICAL THINKING

The Formation of Concepts
in Modern Mathematics

By

FRIEDRICH WAISMANN

New College, Oxford University

With a foreword by
KARL MENGER

HARPER TORCHBOOKS / *The Science Library*

HARPER & BROTHERS · PUBLISHERS · NEW YORK

INTRODUCTION TO MATHEMATICAL THINKING

Copyright 1951 by Frederick Ungar Publishing Co.

Printed in the United States of America

All rights in this book are reserved.
No part of the book may be used or reproduced
in any manner whatsoever without written permission except in the case of brief quotations
embodied in critical articles and reviews. For
information address Harper & Brothers
49 East 33rd Street, New York 16, N. Y.

Reprinted by arrangement with Frederick Ungar Publishing Company; translated from the German, *Einführung in das mathematische Denken,* by Theodore J. Benac.

First HARPER TORCHBOOK edition published 1959

Library of Congress catalog card number: 59-13848

Foreword

Two methods are used to spread scientific knowledge among a larger circle of readers. One kind of popularizing deals with descriptions of the problems, and the external facts that have led to their statement and solution. With the real difficulties of the subject matter, however, it deals as briefly as possible, and at best leaves the reader to surmise its essence through ingenious comparisons. The other method refuses to bypass essential difficulties and even strives to bring those very difficulties closer to the reader. Only one course achieves this goal, and that is absolute clarity—which is frequently not found even in some original articles written for a restricted circle of readers.

To follow this second course to the end makes great demands on the author. Even the most ingenious comparisons and the most brilliant remarks in popular presentations of the first kind fail only too often to remove the doubts of the expert into whose hands they fall as to whether the author himself completely grasps the subject. Rather, absolute clarity is possible only to the scientific writer who has really penetrated the subject under consideration.

To be sure, popular books of the second group also place somewhat higher demands on the reader. Under the guidance of the author, he must reason through many sequences of unusual conceptions which he may at first find a little troublesome. But then he can be sure he will not be led to superficial semi-understanding, but to real insight.

I find it most welcome that in the field of the most modern mathematics the author of this volume has used his pedagogic skill to write a popular presentation introducing the reader to that method of thinking which actually guides the creative scholar in important parts of this science. The reader of this book will find no anecdotes connected with the fringes of scholarly mathematical work; nor will he get a cursory survey of a thousand more or less important problems of interest to the mathematician. He will, however, gain a fundamental insight into the methods of dealing with some very basic questions, above all such that are of interest to the philosopher. This insight will perhaps at first be acquired only with some difficulty. At any rate, because of the clarity of the presentation, this difficulty will be kept to a minimum.

Over and above this, questions of mathematical philosophy are also considered in the text. The author touches upon fields concerning which the most prominent scholars have held sharply divergent opinions even up to the present.

Do mathematical propositions have an empirical origin, as Mill and Mach held? Shall we believe Kant, who declared the arithmetical and geometrical propositions to be synthetic a priori judgments? Was Poincaré correct when he said that the basic rules of arithmetic were certainly synthetic a priori judgments, but that geometrical propositions were analytic—or Frege, who held that the basic rules of arithmetic were analytic and geometrical truths synthetic? Or can we finally follow those who, as in the case of Russell, characterize all mathematical propositions as analytic? Are mathematical propositions vouched for by experience? In the last analysis, do they rest on intuition and experiences of evidence? Are they founded on the fact that mathematics is a part of logic and that the latter, as is frequently said today, is a system of tautologies? Or does the foundation of mathematics rest on the proof of its consistency?

For my part, I believe that none of these questions are to be answered affirmatively. What the mathematician does is nothing but deduce statements with the help of certain methods to be enumerated, and selectable in various ways, from certain statements to be enumerated, and selectable in various ways—and all that mathematics and logic can state about the mathematician's activities which can neither supply nor requires a "foundation," is contained in this simple statement of fact. The basic approach of this book is along similar lines, which also presents new thoughts about some questions of mathematical philosophy. It is quite clear that it may hardly be possible to find undivided assent in this domain, in view of the differences of opinion described above. The reader may find much that is stimulating even on points on which, perhaps, he does not entirely agree with the text.*

Mathematics is used in theoretical physics and in many branches of technical science, recently also in branches of biology and economics. Statements will, therefore, frequently be brought forward in a form so general and concise that it is absolutely necessary thoroughly to understand the practice of the mathematical deduction of statements from statements if one wishes to follow through the formulation of individual statements and

---

* Experts in the philosophy of mathematics will note, as explained in the chapter on complete induction, that—to put it quite simply—its addition to the fundamental propositions or derivation rules permits the derivation of a comprehensive domain of statements which could not be derived without its inclusion, that is, statements about "all" natural numbers, since complete induction is regarded as a convention that regulates the use of the word "all" in the case of natural numbers.

In regard to the definition of numerical equivalence, which goes back to Cantor and Frege, one will willingly grant that it represents only one of many possible attempts at precise formulation of the vague and ambiguous use of this word in colloquial language—a one-sidedness that, to be sure, it shares with every attempt at precise definition of any part of colloquial speech. Furthermore, one will perhaps agree that the criteria given by this definition of numerical equivalence for its application to experience, are none the more precise if compared with those of the usual definitions of fundamental physical concepts, such as equality of lengths, simultaneity, etc. Neither does it seem to me, however, that they are less precise. Indeed, the great mathematical significance of this definition rests preponderantly on its fruitfulness, i.e., on the fact that so many conclusions can be deducted from it. However, just because this definition has proved to be so very fruitful that no other definition up to now has been able to compete, it certainly is of value to point out other similar possibilities (especially to prevent the erroneous view that it is the only conceivable definition).

Foreword

the combination of various statements, and especially if one wishes to proceed from starting propositions to conclusions.

But the knowledge of mathematical methods would also, I believe, be of great value even in sciences where the situation is different, as, for example, jurisprudence, sociology, and those branches of economics in which practice in the various special methods of mathematical deduction is not necessary. Indeed, in practically every discussion on any subject at all there are occasions for making use of the aforementioned insights. This does not mean that by the wider dissemination of insight into the methods of mathematics more intelligent things would necessarily be said than are said today, but surely fewer unintelligent things would be said.

It is rather unimportant which branch of mathematics is studied for the purpose of this theoretical propaedeutic, whether it be arithmetic or algebra, analytical geometry or axiomatics of elementary geometry, set theory or modern logic. What matters is that the book or lecture does not entirely neglect general methodology. Textbooks of logic, totally untouched by the modern development of this science, as they are used today for philosophic propaedeutics, are, to be sure, unsuited for such an introduction to mathematical and logical methods. Indeed, mathematical instruction is so specialized that even in many advanced mathematical textbooks and lectures it neglects the basic viewpoint, from which even the nonmathematician could gain so much profit.

Yet this methodical point of view is precisely what is brought to the forefront in this volume. And therefore wide distribution of this book will surely be of great value in many respects.

Karl Menger

Author's Preface

It is the aim of these reflections to give an insight into the nature of mathematical concept formation, that is, to point out in the activities of the mathematician what might be of interest to a philosophically minded observer. This immediately differentiates the volume from textbooks of mathematics. The reader will not find here a system of theorems with completely developed proofs. He will find no calculations of examples nor applications of mathematics. All that is pushed into the background in favor of a presentation of mathematical ideas.

In the first place, we shall be concerned here in greater detail with the structure of the realm of numbers. The choice of this topic needs a brief justification. Proceeding from intuitive points of view, Leibniz and Newton created differential and integral calculus. In the eighteenth century these investigations soared extraordinarily, one brilliant discovery following another in the sphere of pure analysis as well as in the domain of their applications. This period of mathematics has been compared, not unjustly, with the period of the great discoverers and the heroes of the sea. The mathematicians of that age had the feeling of stepping into a new intellectual world, eager to explore the contours of the continent that sprang up before them out of the mist. In odd contrast to this series of wonderful discoveries was the obscurity that spread over the foundations of the entire concept creation. It cannot be maintained that Leibniz and Newton were very clear about the meaning of a differential quotient. Their expositions vary, but on the whole they had a dim notion of calculating with infinitely small quantities. What this means is difficult to say; and so a certain obscurity has been connected with "infinitesimal calculus" since its birth. Clear-thinking minds, such as the philosopher Berkeley, have not been sparing with their criticism; in the treatise "The Analyst" (1737) the reader finds a very detailed discussion of the new science, a discussion that turned out to be rather crushing. A remark by Lagrange, who lived at the end of the eighteenth century, attests to the fact that it was not only philosophers who had such thoughts but that even the mathematicians did not feel very comfortable in their activities. He found that the condition of mathematics was really deplorable; that it swarmed with contradictions; and that if, in spite of all this, it had achieved such great success, it was only because God in his infinite goodness had ordained that the errors cancel one another out.

No wonder this calculus appeared as something mysterious, almost as something mystical, an art more than a science, prompted by inspiration but not accessible to logical thinking. This view has even infiltrated into textbooks. For example, in Lübsen, an author well known in the last century, we read that differential calculus is a mystical method operating with infinitely small quantities; the differential is a breath, a nothing. Then follows an English quotation: an infinitesimal is the spirit of a departed quantity.

This view has lived on in the general public to the present day, and it has given rise to many an odd idea. An example of this is the well-known book of Vaihinger, "Die Philosophie des Als-Ob (The Philosophy of As-If)," in which the opinion is advanced that our theoretical thinking is frequently guided by fictions, that is, by wittingly false assumptions, which, however, have held up by their results. Vaihinger regarded differential and integral calculus as a mainstay of this view, for he thought that their basic concepts had an entirely fictitious nature. It is significant that all the authors whom Vaihinger summons regarding his view are, in every case, mathematicians of the seventeenth and eighteenth centuries, that is, men who could not have any knowledge of modern ideas.

In reality, the first half of the nineteenth century had already brought some light into this darkness; we mention here only Gauss, Cauchy, and Bolzano. These investigators prepared the way for the new critical period of mathematics, which, much more than previous periods, insists on clear definitions of concepts and logical rigor of proofs. Their work was continued and, in a certain sense, completed by Weierstrass, Cantor, and Dedekind. In the investigations of these scholars it now turned out that the real root of the difficulties lies in a clear comprehension of the concept of continuum. This concept is very closely connected with the concept of irrational number; and so we understand why these investigators were finally led to the examination of the number concept. Since the lectures of Weierstrass, it has been customary to begin a rigorous presentation of differential and integral calculus with a discussion of the number concept.

We, too, will take this course in order to become acquainted with the most important concepts of present-day mathematics.

F. W.

CONTENTS

	Foreword	V
	Author's Preface	IX
1.	The Various Types of Numbers	1
2.	Criticism of the Extension of Numbers	12
3.	Arithmetic and Geometry	18
4.	The Rigorous Construction of the Theory of Integers	25
5.	The Rational Numbers	49
6.	Foundation of the Arithmetic of Natural Numbers	66
7.	Rigorous Construction of Elementary Arithmetic	79
8.	The Principle of Complete Induction	88
9.	Present Status of the Investigation of the Foundations	100
	A. Formalism	100
	B. The Logical School	107
	C. Outlook	116
10.	Limit and Point of Accumulation	123
11.	Operating with Sequences. Differential Quotient	141
12.	Remarkable Curves	153
	Appendix: What Is Geometry?	175
13.	The Real Numbers	182
	A. Cantor's Theory	185
	B. Dedekind's Theory	198
	C. Comparison of the Two Theories	205
	D. Uniqueness of the Real Number System	208
	E. Various Remarks	214
14.	Ultrareal Numbers	219
15.	Complex and Hypercomplex Numbers	226
16.	Inventing or Discovering	235
	Epilogue	245
	Index	247

1. The Various Types of Numbers

The numbers presented to us at the first stage of development are the natural or cardinal numbers 1, 2, 3, 4 . . .; they are used for counting purposes. Numbers are usually represented geometrically as points on a straight line. We will frequently employ this technique to make the following investigations clearer. For this purpose we choose an arbitrary point on a line as the starting point, also an arbitrary interval as the unit of length, and then successively mark off this interval in one direction. The numbers 0, 1, 2, 3 . . . are assigned to the points thereby generated.

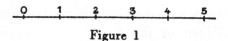

Figure 1

These points are now the "images" of the numbers, and it is advantageous for many purposes to tie our concepts to this scale of points. Henceforth we will speak of the number series also as a point series.

What properties belong to the system of natural numbers?

1. It is an *ordered system*. This means that if two distinct natural numbers are given, one must precede the other; in other words: the relations $a > b$, $a = b$, $a < b$ (a greater than b, a equal to b, a smaller than b) form a complete disjunction.

1. THE VARIOUS TYPES OF NUMBERS

2. Consequently the concept of "betweenness" can be applied to natural numbers; that is, to say that the number c lies between a and b, implies $a > c > b$ or $a < c < b$. On examining the natural numbers with respect to this concept, we encounter a rather characteristic property: every number lies between two others, its immediate predecessor and its immediate consequent. *Between two numbers immediately following one another no further number can be inserted.*

3. There is only one exception. The number 0 *does not have a predecessor*. On the other hand, there is no number which does not have a consequent. We will express these facts as follows: the number series has a first but not a last element; or: it is *infinite on one side*.

The possibility of mapping numbers on the points of a line rests on the fact that the above properties can be ascribed to the point series. Thus it is ordered as soon as we run through the points, let us say, from left to right and think of those points which lie further to the left as antecedent. The other quoted properties also apply. Hence the structure of the number system can be carried over to that of the point system.

We obtain two further properties of the system of natural numbers as soon as we take the arithmetical operations into consideration. Which of the four basic rules of arithmetic (addition, subtraction, multiplication, division) can be performed unlimitedly in this domain, so that the result is always a natural number? Obviously only two — addition and multiplication. In contrast, the subtraction a — b can be carried out only if the minuend a is greater than the subtrahend b; and the division without a remainder only if the dividend is a multiple of the divisor.

If we combine the numbers of the domain arbitrarily by addition and multiplication, we never leave the domain. The natural numbers appear in this respect as a totality, a closed system. We will state these facts as follows: the domain of the

1. THE VARIOUS TYPES OF NUMBERS

natural numbers is closed under addition and multiplication, not closed under the two other arithmetical operations. This latter fact was actually the reason for extending the system of natural numbers in two directions. First we have to introduce the negative and secondly the fractional numbers in order to settle the closure of the number domain. Let us now take a closer look at these number creations.

The negative numbers can be thought of as generated by reversing the formation rule of the number series. Thus, instead of successively adding the number 1, we descend from the number 3 to the number 2, from 2 to 1, from 1 to 0, and then to numbers which we designate sequentially by —1, —2, —3, etc. These are represented by points as indicated below:

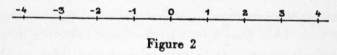

Figure 2

The system of positive and negative numbers is called the system of *integers*.

Let us now compare the integers with the natural numbers. What properties remain intact under this extension? The integers are also an ordered system; therefore the concept "betweenness" has a meaning; however, there is now no longer a number which precedes all others; the system has neither a first nor a last element; it is *infinite on both sides*. Moreover, three operations can be performed unlimitedly — not only addition and multiplication but also subtraction. However, in general, the quotient of two integers is not an integer.

The *fractional numbers* must be introduced if division is to be performed unlimitedly. The system of integers and fractional numbers is called the system of *rational numbers*. This system is closed under all four arithmetical operations. Hence we always stay within the system whenever we combine

1. THE VARIOUS TYPES OF NUMBERS

the individual elements by the four operations. A domain with this property is also called a "field" (the word is used here in a technical sense, say as in "battlefield" or in "field of force"). The system of rational numbers is ordered; for if two unequal fractions are given, one must be greater or smaller than the other. Moreover, we can think of the integers as written formally as fractions. If we try to represent the rational numbers according to the earlier model, as points on a line, we are confronted with a characteristic difficulty. Obviously the fractions lie between the integers, and consequently we must insert further points in the space between the equidistant points of Fig. 2. But how are these points spread out? If we start out, say, from the number $\frac{1}{2}$, is there still an immediate predecessor or an immediate consequent? By no means! For if we choose a fraction which lies as close to $\frac{1}{2}$ as we wish, it is a simple matter to obtain another fraction which lies still closer to $\frac{1}{2}$. (The reader may prove this by showing that the number $\frac{a+c}{b+d}$ always lies between the two rational numbers $\frac{a}{b}$ and $\frac{c}{d}$.)

The totality of rational numbers therefore possesses a structure completely different from that of the natural numbers and integers. Between two rational numbers there always exists another rational number. In order to characterize this specific structure the concept "dense" has been coined. We define an ordered system of elements as dense if between any two elements of the system there always lies another element of the system. The system of rational numbers is our first example of a dense system. The natural numbers and integers do not have this property.

The property of denseness makes it especially difficult to obtain an intuitive picture of the distribution of the rational numbers. We can certainly insert further points between those corresponding to the integers; however this process of insertion must be thought of as continued without end, so that in every

1. THE VARIOUS TYPES OF NUMBERS

interval of the number axis, no matter how small, there will lie an infinite number of rational points. In trying to visualize this system as a completed totality we are inevitably confronted with certain oddities. To illustrate, let us consider the totality of proper fractions with the exception of 0 and 1. The class of these points can obviously be placed in the interval of the number axis between 0 and 1; therefore we will think of it as a point set which covers this interval as an infinitely fine dust. This view leads to the following thought. When I run over the line, say, from the left to the right and beginning from a point to the left of 0, I must at some time or other meet a first element of the point set and when I have run through the whole interval, also a last element; the set must possess a point which is furthest to the left and another which is furthest to the right. A moment's reflection shows that this is absolutely impossible. It follows from the structure of the rational numbers that there is no smallest proper fraction (and also no greatest one). Conceptually this situation presents no difficulty whatsoever; the class of proper fractions is clearly and sharply defined; however the attempt to realize this concept as an intuitively clear picture leads to paradoxes. This illustrates the fact that, though we can learn some things about such relations by graphic methods, we will be misled if we entrust ourselves to them alone.

The stepwise extension of the number domain, which we have sketched above, is brought to some kind of a close with the rational numbers. And many will be tempted to assume that the extension can be carried no further, since the rational points fill up the number axis completely and without holes. That this is a mistake, that the rational numbers, even though sown infinitely dense, still do not cover the entire number axis, is the great discovery of Pythagoras. He first recognized that there are numbers which, though completely different from the rational numbers, are still related to them. We will illustrate this extraordinary discovery by constructing a square

1. THE VARIOUS TYPES OF NUMBERS

on an interval of length 1 and drawing a diagonal in this square. Our intuition tells us that this diagonal must have a very definite length. Let us try to compute it. According to the theorem of Pythagoras the square of the diagonal is equal to the sum of the squares of the two legs of the right triangle, therefore 2. Consequently the diagonal has a length $\sqrt{2}$; $\sqrt{2}$ is that number whose square is 2. Can it be represented as a fraction? First it is clear that it lies between 1 and 2; therefore we experiment with $1\frac{1}{2}$; the square of this number is $\frac{9}{4}$, therefore it is too large. The number under investigation must accordingly be greater than 1 but less than $1\frac{1}{2}$; $\frac{4}{3}$ is such a number; the test shows however that it is too small. Let us continue this process of inserting a fraction between those hitherto investigated and testing whether its square is exactly 2. In this way we will find numbers which are either too large or too small, and which we can arrange in the form of two sequences:

too small:	too large:
1	2
$4/3$	$3/2$
$7/5$	$10/7$
$24/17$	$17/12$

Now one could think that if we continue this search on and on and take all the time and effort that is needed, we must eventually arrive at a number whose square is exactly 2. This point will be clarified as soon as we have determined whether the trials attempted so far have failed due to chance or because there is a deeper reason underlying it. If there is a rational number which is exactly $\sqrt{2}$, then there is a fraction $\frac{p}{q}$ such that $\frac{p^2}{q^2} = 2$. Now a fraction is equal to an integer only if the denominator goes into the numerator without a remainder. Hence p^2 is divisible by q^2. But this is only possible if p is also divisible by q. For, if p and q are two relatively prime numbers (i.e., two numbers which have no common prime factors), then p^2 and q^2 are also relatively prime; by the squaring process no prime factors can be generated which do not already exist. Hence if $\frac{p^2}{q^2} = 2$, that is, an integer, then $\frac{p}{q}$ must also be an integer. However this is impossible since $\frac{p}{q}$ lies between 1 and 2, an interval in which there are no integers.

Thus the simplest reflection on the divisibility properties of numbers readily shows that the attempt to find a rational number whose square is exactly 2 must be fruitless. On the other hand there can be no doubt that the diagonal of the unit square has a very definite length. If we think of this length as laid on the number axis with one end at 0, we obtain a point which is the geometrical representative of $\sqrt{2}$; the point where $\sqrt{2}$ lies can not be a rational point of the number axis. Hence we have the following result: Even though the rational points cover the number axis as an infinitely fine

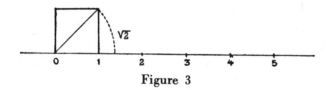

Figure 3

dust, they still do not completely fill it up. They form, as it were, a porous system, in which cracks and crevices leave room for another type of number, the *irrationals*.

What can we say about the distribution of the irrational numbers? That is, are they exceptions — to be found between the rational numbers only here and there? We can answer this question by a very simple argument. Let us think of the entire number axis with the rational points on it as enlarged in the ratio $1 : \sqrt{2}$, that is, in such a way that $\sqrt{2}$ is used as unit interval. Then every rational number will go over to a number which can be shown, as in the case of $\sqrt{2}$,

7

1. THE VARIOUS TYPES OF NUMBERS

to be irrational; for instance, 1 in $\sqrt{2}$, $\frac{3}{10}$ in $\frac{3}{10}\sqrt{2}$, etc. We thereby obtain a second system which is also dense, consists only of irrational numbers, and is somehow squeezed in between the rational numbers. But irrational numbers can be generated in many ways; for instance not only by square roots but also by cube roots, fourth roots, etc. In fact there are infinitely many operations which will produce irrational results when "set loose" on an individual rational number. These remarks lead us to surmise that the irrational numbers make up the principal part of the structure of the number axis instead of being its exceptional points. In the theory of sets it is actually shown that most of the points on the number axis are irrational and that the rational numbers are vanishing exceptions.

The system of rational and irrational numbers is called the system of *real numbers*.

We can also look at these relations from another point of view by proceeding from the representation of numbers as decimal fractions. A decimal fraction can terminate or continue without end. We will state, first of all, that every terminating decimal fraction can be transformed into a non-terminating one by diminishing the last numeral by unity and permitting only nines to follow afterwards. For instance, we have $\frac{1}{2} = 0.5 = 0.4999 \ldots$

If we use this property, we can represent the totality of the real numbers by non-terminating decimal fractions. These fall into two categories: the periodic and non-periodic decimal fractions, the former corresponding to the rational numbers, the latter to the irrationals. (We omit the proof, which is simple.) This shows anew that between two rationals there must always lie irrational numbers; for we can always insert arbitrarily many non-periodic decimal fractions between two periodic ones.

By passing on to the imaginary numbers we take a new direction in the extension of the number concept. They were unknown in antiquity. They appeared for the first time in a work of Cardano (1545), whose name is connected with the solution of cubic equations. However, the mathematicians of that day did not have a clear understanding of the nature of these quantities. On the contrary, the imaginary numbers forced themselves into calculations against the desires and inclinations of mathematicians. This situation resulted from algorithmic requirements. Cardano's formula often represents the solution of a cubic equation, even though it may be real, in a form in which square roots of negative numbers must be extracted. Now, there is no real number whose square is negative; therefore these roots are "impossible." However, in spite of these scruples this new type of expression was manipulated like an ordinary root; and the end justified the means. We encounter here a factor which on the whole played an important role in the history of mathematics. It seems that an independent, onward-driving force is inherent in the manipulation of formulae, in the algorithms; and that in our case it induced the mathematicians to handle imaginary numbers; to the great advantage of mathematics; for the pedantic requirements of rigor would probably have paralyzed the further development. Fortunately the mathematician of that day treated subtle logical concepts with indifference; not, however, to such an extent that he would not retain, when operating with these remarkable entities, a certain uneasiness, a bad conscience, which betrayed itself in names as "impossible" or "fictitious numbers." As evidence of this, Leibniz made the following statement in the year 1702: "The imaginary numbers are a fine and wonderful refuge of the Divine Spirit, almost an amphibian between being and non-being." We note here a reflection of the strange impression which these numbers must have made on the mathematician. Thus we find that Euler was candidly astonished by the remarkable fact that a number as $\sqrt{-1}$ is neither smaller nor greater than 5, neither positive nor negative, and that it cannot be compared

1. THE VARIOUS TYPES OF NUMBERS

with ordinary numbers at all. And when a student hears about imaginary numbers for the first time, he again experiences this impression of mysteriousness, which disappears later in proportion as he learns to use these numbers. However, the nature of these numbers is not made clearer by usage. We simply have to become accustomed to them and ask nothing further. Under these circumstances it marks an epoch-making development that Gauss should give a geometrical representation of the imaginary numbers. It is found for the first time in his own abstract of a number-theoretical work in the year 1831 and has made an extraordinary deep impression. However, we know from Gauss' diary, which was left among his papers, that he was already in possession of this interpretation by 1797. Through this representation Gauss intended to clarify the "true metaphysics of imaginary numbers" and bestow on them complete franchise in mathematics. Now what is this representation? We already know that the rational and irrational numbers fill up the number axis, so that there remains not even the smallest of gaps. Hence, if we now wish to interpret the imaginary numbers geometrically, we must use a second line. In the Gaussian interpretation the real numbers are represented as points on the x-axis, the imaginaries as points on the y-axis of a rectangular Cartesian system of coordinates, whose intersection point represents the number 0. Hence a rotation through 90° takes the positive real number axis into the positive imaginary number axis. Gauss did not give a basis for this representation; however he derived from it the right to operate with imaginary numbers.

By means of this interpretation we can also obtain a geometrical picture of those numbers which are generated by the addition of an imaginary and a real number, as $2 + 3i$, the so-called *complex numbers*. Such a number can be represented as the point ($x = 2$, $y = 3$) of the coordinate plane. We thereby see that the images of the complex numbers are distributed over the plane. For the representation of complex numbers

1. THE VARIOUS TYPES OF NUMBERS

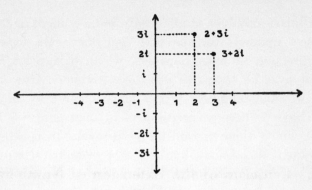

Figure 4

a line no longer suffices. We must have recourse to a plane; the number world has been broadened to a two-dimensional manifold. This clearly shows the reader that a far-reaching step has been taken. Up to now if one has understood by a number something which could be arranged serially by "greater" and "smaller," then this is no longer valid in the domain of complex numbers. For example, which of the two numbers $2 + 3i$ and $3 + 2i$ is the greater? The (linear) order is not valid and therefore neither is the concept of "betweenness." This shows that in the transition to complex numbers it is no longer possible to compare numbers with respect to their magnitude, a property which hitherto was thought of as wholly essential for the concept of number.

2. Criticism of the Extension of Numbers

In the previous chapter we presented an introductory, orienting view of the extensions of the number domain in order to acquaint ourselves as soon as possible with the subject matter under investigation. In particular we apprehended that the reason for introducing new numbers was the requirement that subtraction, division and evolution be operations which could be performed in all cases.

This is the way in which our subject matter is usually represented, as it is studied today in school and, generally speaking, as things developed historically. And yet it is easy to raise questions which perplex us from this point of view. Can we continue the extension of the number domain still further? Can we invent numbers which are no longer representable in the plane, but which require a mapping in the three-dimensional space? Or is this impossible? And on what does it essentially depend? However there is a question which is more important than these. Up to now we have said that it was the desire to make certain operations possible without any exception, which urged us to extend the number domain. Thus it was the non-applicability of subtraction which led to the introduction of negative numbers; while that of extracting roots, led to the introduction of the irrationals and later of the imaginary numbers. We could describe this as follows: the proposition to subtract 7 from 5 has no solution only as long as we restrict ourselves to the natural numbers; this does

2. Criticism of the Extension of Numbers

not prove that it does not have a solution in the absolute sense, but only that the domain of natural numbers is too meager to enable us to solve the proposition. Consequently we extend this number domain by annexing the negative numbers, and now the solution exists. But is this always possible? Could we solve every insolvable problem by introducing new numbers and writing the solution in terms of a new number? For instance, the operation $\frac{1}{0}$ has no solution in the usual arithmetic; for there is no number which gives 1 when multiplied by 0. Could we in this case, too, argue that only the present numbers are insufficient to solve this problem? If this point of view is valid, let us extend the domain of numbers to include a solution of this operation. Let us set $\frac{1}{0} = \omega$ and then calculate with ω as we did earlier with i. Well now, let us try to build a new arithmetic on this foundation. Is this legitimate?—The equation $1^x = 2$ has a solution neither in the domain of the real nor of the complex numbers. Could we not force the solvability by declaring that certainly there is a number which satisfies this equation? Let us consider the following two equations:

$$x + y = 10 \qquad 2x + 2y = 30$$

Everyone will say that these equations do not have a solution, for the second contradicts the first. Shall we reply that this is true but only when we restrict ourselves to the numbers known up to now? What is to hinder us from inventing a new kind of number by which such a system of equations can be solved? Nothing—if the introduction of new numbers only amounts to postulating the existence of numbers which solve the stated problem. But is this actually a legitimate procedure? We certainly should not confuse wishful thinking with wish fulfillment. Consequently to wish that a number shall exist whose square is 2 or whose square is -1 is not the same as saying that such a number actually exists. "Why not also ask

2. Criticism of the Extension of Numbers

that a line pass through three arbitrary points? Because this condition contains a contradiction. Above all one must first prove that these other conditions do not contain contradictions. Before one has done this, all striving after rigor is nothing but mere pretense and sham." (Frege.) And Russell remarked, "the method, whereby one 'postulates' what one needs, has many advantages. They are the same as the advantages of the thief face to face with an honest task." No, in the postulational method there does not reside a secret magic power. It is an expedient which is far too evasive to be valid.

Hence we must admit that the entire present structure of the number world hangs in the air; we do not quite know whether the negative, the fractional, the irrational numbers exist;[1] we do not even know what permits us to extend the number domain. We must begin anew.

However, perhaps our criticism goes too far. An advocate of the customary interpretation could contend that entities such as negative, fractional, irrational numbers clearly exist in the various applications of arithmetic. For instance, the existence of $\sqrt{2}$, that is, the solvability of the equation $x^2 - 2 = 0$ follows very conclusively from the interpretation of Pythagoras, whereby the diagonal of the unit square has the length $\sqrt{2}$. Similarly, the existence of fractional numbers can also be established from a purely geometrical point of view, by dividing the unit interval into equal parts. And in the case of negative numbers we not only have their representation on the number axis but also their application to hot and cold temperatures, to assets and debits, to elevation above and depression below sea level, and so on; therefore the calculation with negative numbers has a clear meaning. Furthermore, such associations have helped the mathematician very effectively in the conception of those new ideas—therefore why should we reject this procedure?

---

[1] Regarding the sense of the existence, cf. the last Chapter.

2. Criticism of the Extension of Numbers

Two points must be considered in order to estimate correctly the value of such opinions.

First, it is reasonable to require that arithmetic should be separated from its applications. There is no doubt that trains of thoughts like those mentioned above are highly suggestive, for they certainly have guided mathematicians in the conception of their ideas. Here, however, we are only concerned with the *justification* of operating with the new numbers, and from this point of view we must admit that the quoted examples are not convincing. Will anyone seriously assert that the existence of negative numbers is guaranteed by the fact that there exist in the world hot and cold, assets and debits? Shall we refer to these things in the structure of arithmetic? Who does not see that thereby an entirely foreign element enters into arithmetic, which endangers the pureness and clarity of its concepts? Even if there did not exist in the empirical world a distinction between hot and cold, assets and debits, this would not affect the right to introduce positive and negative numbers. If we were to base the existence of these numbers on such facts, then we would be tying arithmetic too closely to the accidental occurrences of the empirical world. And finally, we do not even feel satisfied by such a representation, since it is not sufficient for establishing the arithmetic of integers. This is substantiated by the fact that, in the attempt to introduce the negative numbers intuitively, the rule of signs, minus times minus = plus, forms the stumbling block; it cannot be made graphic by such a demonstration, therefore, such attempts complicate rather than clarify matters. And what shall we say about certain higher complex numbers, certain transfinite numbers, as introduced by Georg Cantor and others, which are capable of no such demonstration? Is this a reason to believe that they don't exist? Or must we first wait until someone has found an application of these numbers to things or events in reality?

2. Criticism of the Extension of Numbers

Contrariwise, the requirement which states that arithmetic should find its hypotheses in itself and refer in no way whatsoever to those things that are not arithmetical—to experience, to perception, or anything else—is certainly appropriate. H. Hankel rejected every attempt of this kind with these clear words: "The condition for erecting a universal arithmetic is therefore a purely intellectual mathematics, one detached from all perceptions, a pure theory of forms, in which it is not quantity or its representatives, numbers, that are tied together, but intellectual objects, concepts, to which actual objects or relations may or may not correspond." (*Theorie der komplexen Zahlensysteme*, Theory of Complex Number Systems p. 10.)

The second remark is that even if we were to refer to geometry when introducing the irrational numbers, we could thereby only recognize the existence of those numbers which could be *constructed*. Now the concept "*constructibility*" can be classified into: constructible with ruler, with compass, with ruler and compass, or with some other kind of mechanism. This means that this concept is always to be understood relative to a set of allowable construction methods. According to the usual way of defining these methods it turns out that the points thereby constructed form only an insignificantly small minority. For instance, with ruler and compass all points on the number axis which correspond to the so-called transcendental numbers (these are numbers which do not satisfy an algebraic equation, for example, π, $\log 2$, $2^{\sqrt{2}}$), cannot be constructed; this is also true of the majority of the irrational numbers which satisfy an algebraic equation—the so-called algebraic numbers. All these points would therefore elude us if we were to make use of such a geometric construction.

However, even for reasons of principle alone, it is not feasible to draw on space perceptions for the foundation of arithmetic. We will try to describe the reasons which forced

2. Criticism of the Extension of Numbers

this position on the mathematician. For this purpose we must take as general a view as possible of the relations between arithmetic and geometry. We begin with a brief description of the structure and development of geometry.

3. Arithmetic and Geometry

Geometry was first developed according to rigorous scientific principles in the στοιχεῖα of Euclid. This treatise was long recognized as a model. It starts with a few intuitively given propositions whose validity is not subject to further discussion; it then seeks to build up the structure of geometry so that all theorems follow by rigorous logical methods from these basic propositions without any further reference to intuition.[1] These basic propositions, which are assumed to be self-evident, are called *axioms*. Among the axioms of Euclid two groups can be distinguished:

1. General axioms of magnitude (Κοιναὶ ἔννοιαι), such as: "Two quantities equal to a third, are equal to one another," "Equals added to equals give equals," "The part is smaller than the whole."

2. The essentially geometric axioms (αἰτήματα). As such Euclid laid down five propositions:
    1. Every point can be joined to any other point by a straight line.
    2. Every straight line can be extended beyond each of its endpoints.

---

[1] Actually, it falls short of this goal of Euclid. Only in modern times, due to the investigations of Pasch, Veronese, Hilbert, etc., has the structure of geometry taken a form sufficiently rigorous so that all proofs can actually be developed in a purely logical fashion without having recourse to intuition.

3. ARITHMETIC AND GEOMETRY

3. A circle can be drawn with an arbitrary radius about any point.
4. All right angles are equal to one another.

And now comes a very remarkable proposition:

5. If two lines are cut by a third so that the angles inside the two lines and on the same side of the third have a sum less than two right angles, then the two lines intersect on the specified side when sufficiently extended.

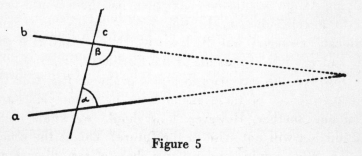

Figure 5

This last axiom, the parallel axiom, was above all the starting point of the historical development. In the presence of such a complicated proposition, the question was naturally raised as to the source of its certainty, for it is too involved to be spoken of as an immediate result of intuition. The validity of this axiom has been doubted since the advent of Euclidean geometry. Consequently mathematicians of later periods tried to derive the parallel axiom from the other axioms and in this way to cleanse geometry of this "blemish." However, these efforts were without success. This led them to try to find an indirect proof of the parallel axiom. For this purpose they assumed that this axiom was false and then derived the consequences of this assumption in the secret hope that a contradiction would eventually be met. Naturally a contradiction would prove that the assumption was not admissible, and that therefore the opposite statement was valid. In these investigations they encountered unusual results. For example, similar figures

3. Arithmetic and Geometry

are impossible if the parallel axiom is not valid (Wallis); also, in the plane there must be a triangle of maximum area. Lambert found that if the parallel axiom was not valid, there would be a unit of length distinguished by nature; hence it would no longer be true that all line segments could be treated in the same way. No matter how remarkable or absurd these results sound, they do not imply a logical contradiction. Thereupon Johann Bolyai and Lobachevski gave the question a wholly new turn. They developed the inferences which resulted from dropping Axiom 5 with the conviction that these would never lead to a contradiction. In this way a totally new "anti-Euclidean" geometry was disclosed which was consistent with itself and was just as possible as the Euclidean.

However, this gave rise to a new problem. It is true that the consequences of these assumptions have not yet contradicted one another. However, how should we know that a contradiction will not arise in the future? This is the crucial question. What would happen if the chain of reasoning should some day produce a contradiction? This would mean that the entire structure of non-Euclidean geometry as constructed up to that time would collapse. Consequently, a doubt hung over the concept creations of Bolyai and Lobachevski like a threatening cloud. This was the first time that the problem of consistency appeared on the horizon of the mathematician.

The course followed by mathematicians of the succeeding generations is of tremendous fundamental interest. Obviously a direct proof of consistency was not likely to be found. For such a proof, strictly speaking, the entire immense chain of conclusions would have to be submitted and surveyed as a complete whole. Instead of this an indirect approach was used. F. Klein found in 1870 that the entire system of concepts of non-Euclidean geometry could be "mapped" on the system of concepts of Euclidean geometry, so that every contradiction of one system would show itself as a contradiction in the other system. In other words, to every concept of non-Euclidean

3. ARITHMETIC AND GEOMETRY

geometry there is associated by means of a definite rule a concept of Euclidean geometry called its image; similarly, for every proposition of one of the theories there is a proposition in the other such that corresponding propositions have the same logical form. If we now replace the basic concepts of non-Euclidean geometry by the corresponding concepts of the Euclidean, then the totality of axioms of non-Euclidean geometry goes over into propositions of Euclidean geometry. We have produced within the framework of Euclidean geometry a "model" for the non-Euclidean. By the above substitution all logical relations between the propositions remain valid. If in the theory T the proposition p is a logical consequence of the propositions q and r, then the same is valid for the corresponding propositions p', q', r' in the theory T'. All inferences within the one theory are transferable without change to the other. Hence, if the axioms of the theory T should actually lead to a contradiction (that is, if two chains of inferences could be given which, starting from the axioms, led in one case to a certain proposition p, and in the other to the negation of this proposition), this must also be true for the corresponding statements in the theory T'.

What this means in individual cases we cannot go into here. We shall, however, indicate at least the first steps of this process. For this purpose, let us think of a fixed circle k in the Euclidean plane. We now set up a lexicon by which the basic concepts of non-Euclidean geometry are associated with certain concepts of the Euclidean:

By a "point" we understand a point in the interior of k.

By a "straight line" we understand that part of a straight line which falls in the interior of k.

Special rules are given for the measurement of angles and distances, whose discussion here would lead us too far afield. Let us only say that those conditions are chosen so that an arbitrary length can be laid off on a line infinitely often, one after another, without leaving the interior of the circle.

3. Arithmetic and Geometry

Naturally this length, if measured by Euclidean standards, becomes smaller and smaller. Intuitively speaking, this means that as a being moves from the center of the circle towards the circumference it shrivels up more and more, and takes smaller and smaller steps, so that it can never reach the edge of the circle, which appears to be infinitely far away. (However, we wish to emphasize strongly that this simile has nothing to do with the demonstrativeness of any part of the consideration; the simile can be abandoned and the subject matter developed from a purely abstract point of view; therefore any controversy regarding this simile is futile.)

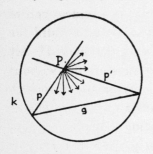

Figure 6.

It then turns out that the structure just defined satisfies all the axioms of Euclidean geometry except the parallel axiom. For, if we consider the lines passing through a fixed point P, they split into two classes, relative to their position to a fixed line g (which does not pass through P); those which cut the line g, and those which do not cut g. These two classes are separated by the two lines p and p'. We shall call these two lines "parallels," since they cut the line g, from the non-Euclidean point of view, only at infinity. Hence through a point not on a line g two lines can always be drawn which are parallel to g; in other words, the Euclidean Axiom 5 is no longer valid. Yet all propositions of non-Euclidean geometry can be interpreted so that they say something about configurations in the interior of a fixed circle k. Consequently a contradiction can certainly not appear if we assume that Euclidean geometry is free from contradiction. It is evident that this is not an absolute proof of consistency. However, this represents a far-reaching step for the question of consistency is removed from a region which is unknown and difficult to envision, to one which is known and easily accessible.

3. ARITHMETIC AND GEOMETRY

However, mathematicians did not stop here. Instead they immediately raised this question: What guarantees the consistency of Euclidean geometry? (For it was just the rise of non-Euclidean geometry and certain other developments, of which we will speak further on, that has deeply affected the naive confidence in intuition.) A way out of this difficulty offered itself to the mathematicians which appears today as the ultimate development of the idea of analytical geometry. The creation of coordinate geometry itself is associated with the names of Descartes and Fermat. It enables us to transfer all geometric (spacelike) relations to a world of pure abstract number-relations. It associates in the well-known manner two real numbers to every point of the plane. These numbers are called the abscissa x and the ordinate y of the point. Every straight line is then given as the totality of all points x, y which satisfy a linear equation $ax + by = c$. Similarly, to every point of space there is associated a triplet of real numbers. Along with the axiomatic structure of Euclid there appeared during the seventeenth century a new structure of an analytic nature. In it, points, lines, and planes are represented by triplets of real numbers, by linear equations and systems of equations, and the relations between the geometrical concepts are drawn from relations between the corresponding arithmetical structures. This arithmetization was carried through in detail by Study; he defined the points of the plane outright as pairs of real numbers, lines as linear equations, etc., and comprehended analytical geometry as a pure computational treatment of some kind of arithmetical model, without thereby appealing to intuition.

Now Hilbert used the above technique to return the question of the consistency of Euclidean geometry to the analogous question for the domain of analysis (the theory of real numbers). He noted that the entire Euclidean geometry is realized in analysis, in other words, that a system of entities can be

23

3. ARITHMETIC AND GEOMETRY

constructed from real numbers which satisfies all axioms of Euclidean geometry under a suitable assignment of names.

From this we can draw the following important conclusion. If a hidden contradiction existed in Euclidean geometry, then it must already be present in the theory of real numbers. Accordingly, it is in the theory of real numbers that we must look for the logical foundation of the other mathematical systems of concepts, especially of the different geometries. Hence it appears intolerable—and thereby we return to the remarks made at the end of the previous section—that, in building up the real numbers, we should continue to have recourse to spatial perceptions, to the very same intuition that had been excluded, as being of suspicious character, from the rigorous construction of geometry. If the theory of real numbers is to support geometry, how can one draw the existence of irrational numbers from geometry? What actually supports the structure, and what is being supported? To have recourse to geometry in the foundation of arithmetic would be to follow a vicious circle, a course that the mathematician wishes to avoid at all cost.

4. The Rigorous Construction of the Theory of Integers

We are therefore confronted with this problem: How shall the various types of numbers be introduced without recourse to geometrical considerations? How can arithmetic be founded in a purely arithmetical manner?

The course followed by mathematicians to find an answer to these questions will first be exemplified by the concept of power. By a power we understand a product which is formed from factors that are all equal.

For example, we have

$$a \cdot a = a^2$$
$$a \cdot a \cdot a = a^3$$

For these powers it is easy to prove the propositions

$$a^m \cdot a^n = a^{m+n}$$
$$a^m : a^n = a^{m-n}$$
$$(a^m)^n = a^{mn}$$

What shall we now understand by an expression such as a^0? The previous definition fails us, since it is meaningless to set the number a zero times as a factor. In the textbooks we frequently find "proofs" that $a^0 = 1$. Such a proof would look somewhat as follows:

Since $a^m \cdot a^n = a^{m+n}$, this must also be valid if we set $n = 0$. This gives $a^m \cdot a^0 = a^{m+0}$, that is, $a^m \cdot a^0 = a^m$. Consequently $a^0 = 1$. This argument is open to criticism since the

4. The Rigorous Construction of the Theory of Integers

generic equation on which the proof rests has actually been proved only if m and n are integers distinct from 0; therefore we have no right to assume the validity of this equation if $n = 0$. In the first place a^0 has not even been defined, for the definition of the power a^n applies only if the numbers n are greater than 0. We can, however, supplement the definition arbitrarily to take care of the case $n = 0$, and now we do so by stating that $a^0 = 1$ by definition. Hence this equation is not one which has to be proved; every attempt to produce such a proof degenerates into a vicious circle.

Now the mind struggles against accepting our assertion that this is a mere convention. It says that there is nothing arbitrary about this equation; instead, there must be a reason for saying that a^0 is exactly $= 1$, and not equal to another number. Hence we will consider for a moment what would happen if we were to make another convention. Let us assume that $a^0 = 5$. Is this possible? Certainly; for this equation merely implies the rule that the symbol „a^0" is always to be replaced by the symbol "5". Why should we not be allowed to do this? If I resolve that in the future I will always write 5 instead of a^0, then I can never contradict myself; this means that there is here nothing to refute, since I am actually proposing a new convention. But it is an entirely different question whether this convention would be useful. The rule $a^m \cdot a^n = a^{m+n}$ would not be valid if $n = 0$. For this case a new rule would have to be devised which has not the slightest connection with the earlier one. Now the mathematician strives to form his concepts so that the rules remain valid as far as possible. If I were to set $a^0 = 5$, this aim would be thwarted. The simplicity and purity of the system of rules would thereby be destroyed. On the contrary, if I set $a^0 = 1$, then the power rules, which had been originally proved only for $m, n > 0$, remain in effect also for this new case. Hence this convention is significantly superior to the other possibilities. And now we recognize the main reason for this convention. Among the

4. THE RIGOROUS CONSTRUCTION OF THE THEORY OF INTEGERS

infinitely many conventions which I can make there is one that is preferable inasmuch as it leaves the calculating rules unchanged. *Hence the preservation of the calculating rules governs the concept formation.* This is only an example of something which H. Hankel has called the *principle of the permanence of the calculating rules* and which we can formulate as follows: If we wish to extend a concept in mathematics beyond its original definition, then among all the possible directions of this extension the one is to be chosen that will leave the calculating rules intact as far as possible. This principle of permanence is not a statement whose validity may be questioned; instead it is, so to speak, a guiding principle for the formation of concepts.

We shall now use this principle to introduce the system of integers in a rigorous manner. We shall extend the concept of difference just as we have extended the concept of power. For this purpose we make a list of the rules for operating with differences in the domain of the natural numbers.

First it is clear that the sum, the difference, and the product of two differences can again be represented as a difference. Thus we have

(1) $\qquad (a-b) + (c-d) = (a+c) - (b+d)$
(2) $\qquad (a-b) - (c-d) = (a+d) - (b+c)$
(3) $\qquad (a-b) \cdot (c-d) = (ac+bd) - (ad+bc)$

These three formulae can be proved to be valid within the arithmetic of natural numbers, as we shall see later. Along with these rules we set three more relations which state that differences can be ordered:

(4) $\qquad a-b = a'-b'$, if $a+b' = a'+b$,
(5) $\qquad a-b > a'-b'$, if $a+b' > a'+b$,
(6) $\qquad a-b < a'-b'$, if $a+b' < a'+b$.

These rules are also demonstrable within the arithmetic of natural numbers. Our objective is to obtain a criterion for com-

4. THE RIGOROUS CONSTRUCTION OF THE THEORY OF INTEGERS

paring the magnitude of two differences in which further differences do not appear.

This shows that we can erect a calculus[1] with differences of natural numbers. However, this calculus is tied to the assumption that the differences are natural numbers; by definition (cf. p. 86) this is true only if the minuend is greater than the subtrahend. Hence our six formulae are demonstrable only under this condition.

We shall now set this limitation aside and erect a general calculus of differences, guided by the principle of permanence. Accordingly, we shall look upon these six formulae as defining what is meant by the concept of "difference" in the extended sense of the word. It will be shown that differences in this new sense are exactly those entities which we usually call positive and negative numbers.

Thereby a very decisive change of course has taken place. We no longer postulate that subtraction can always be performed; we waive the hypothesis that there are negative numbers. Instead we construct a system of concept objects which we endow, on the authority of arbitrary conventions, with those properties which make them behave as if they were positive and negative numbers. We call these concept objects integers.

The concept objects we are speaking about are the differences of two natural numbers. However, in order to avoid even the appearance of assuming that subtraction can be performed unlimitedly, we shall plan the entire calculus so that the concept of difference will not be mentioned at all.

Instead of differences we consider number couples (a, b). Everyone will grant that we can put two arbitrary natural numbers together as a couple. This does not involve a postulate, and it is all that we need to construct the following calculus.

[1] Translator's note: the word "calculus" is used here in the wide sense, namely, to indicate a method of calculation; this is at variance with the common usage whereby the unmodified word stands for "infinitesmal calculus."

4. THE RIGOROUS CONSTRUCTION OF THE THEORY OF INTEGERS

The only basis on which we shall erect this structure will be the system of natural numbers and the calculating rules which are valid in this domain.

The objects of our consideration are pairs of natural numbers. Here the couple (a, b) will be distinguished from the couple (b, a). At present we actually do not know what such a number couple "stands for" and attach no intuitive meaning to it. It is only by giving the rules of application for these symbols that we build up its meaning.

1. *Definition of equality*

We first define when two number couples shall be called "equal." Our definition states
$$(a, b) = (a', b'), \text{ if } a + b' = a' + b.$$
According to this definition we can test the equality of two number couples by examining whether the number expressions standing on the right are equal or not. Thereby the equality of number couples is reduced to the equality of natural numbers. It will also be true in general that we will convert every statement about number couples to one which involves only relations between natural numbers. Consequently the entire arithmetic of number couples only represents a new way of speaking which can be translated into the language of ordinary arithmetic. If someone were to ask: "Are these number couples actually equal?" we would have to reply that the word "actually" is illegitimate here. At the present we do not even know what the word "equal" means. It will only be by laying down a criterion for equality that this term will become meaningful; and how we choose the criterion stands at our discretion.

However, are we really free in the choice of a criterion? Would any convention which could occur to us have served the same purpose? To this we reply that the concept "equal"

4. THE RIGOROUS CONSTRUCTION OF THE THEORY OF INTEGERS

already has a very definite sense in the arithmetic of natural numbers. Consequently, in dealing with statements about the equality of two number couples, our procedure will be justified[2] only if equality in the system of number couples plays a role entirely analogous to that of equality in the system of natural numbers; or, to state this more clearly, if the concept of equality in the new thought domain exhibits the same formal properties as the concept of equality in the old. Now what are these properties?

Equality is a relation between natural numbers which is
1. reflexive, that is, $a = a$;
2. symmetric: if $a = b$, then $b = a$;
3. transitive: if $a = b$ and $b = c$, then $a = c$.

Consequently, if a definition of equality is set up in the domain of number couples or, for that matter, in any other part of mathematics, we are obliged to prove[2] that it has these three formal properties. Of course a definite relation would still be defined even though it did not satisfy these properties. However, why should we designate such a relation as equality? It would have little in common with the usual concept of equality, and it would be misleading to use the same word. Hence the choice of a definition for the equality of number couples is restricted by these conditions. Does the above definition meet these requirements?

As to 1. It is easily seen that every number couple is equal to itself. For by definition $(a, b) = (a, b)$ means that $a + b = a + b$, and this is valid no matter what values are substituted for a and b.

As to 2. It must be shown that the equation $(a, b) = (a', b')$ leads to the equation $(a', b') = (a, b)$.

The first equation means: $a + b' = a' + b$;
the second equation means: $a' + b = a + b'$.

The second equation follows from the first by interchanging the sides.

[2] Cf. the remark of p. 63.

4. The Rigorous Construction of the Theory of Integers

As to 3. From $(a, b) = (a', b')$
and $(a', b') = (a'', b'')$
follows $(a, b) = (a'', b'')$.

For the proof we replace the equations between the number couples by equations between numbers. We must then show that from
$$a + b' = a' + b$$
and $a' + b'' = a'' + b'$ follows $a + b'' = a'' + b$.
If we add the first two equations, we obtain
$$a + b' + a' + b'' = a' + b + a'' + b',$$
and on cancelling the terms a' and b',
$$a + b'' = a'' + b,$$
which is exactly the equation to be proved.

Thereby it is clear that the concept of equality actually has the required properties and that our definition was legitimate.

Many readers will think that this entire derivation is superfluous. For, doesn't the concept of equality imply that every object is equal to itself? Of course this is the meaning of equality. But, who is going to tell us that the relation introduced in definition (1) really has the properties of equality? This must first be proved. For instance, if we had defined:
$$(a, b) = (a', b') \text{ if } a + b' = 2a' + b,$$
this would also have been a definition of the sign "=". In this case, however, a number couple need not be equal to itself. This example shows that the fulfillment of this condition is not to be taken for granted.

Before proceeding we will draw an important inference from the definition of equality. It is
$$(a, b) = (a + c, b + c),$$
as one recognizes immediately from the definition (1). This means that a number couple remains unchanged on adding the same number to the antecedent and the consequent. On reading the equation from right to left, it tells us that a number couple also remains unchanged if the same number is taken away from each term. Consequently a number couple can be

represented in an infinite number of forms. For instance, we have
$$(5, 3) = (6, 4) = (3, 1) = (2, 0) = (9, 7) = \ldots$$
Among all these forms there is one of particular interest, namely, (2, 0). It is now obvious that there are three possibilities for a number couple:

1. The antecedent is greater than the consequent; in this case
$$(a, b) = (a - b, 0).$$

2. The antecedent is smaller than the consequent; in this case
$$(a, b) = (0, b - a).$$

3. The antecedent and consequent are equal; in this case
$$(a, a) = (0, 0).$$

In the first case we say that the number couple is *positive*, in the second case *negative*, in the third case a *null couple*. Every number couple belongs to one and only one of these three categories.

2. *Definition of "greater"*

We say that a number couple is greater than another, in symbols $(a, b) > (a', b')$, if $a + b' > a' + b$.

Again we must test whether this definition is justified. This means whether the concept "greater" just formulated has the formal properties which belong to it in ordinary arithmetic. In the latter the concept "greater" is

1. irreflexive, which means that a is never greater than a,
2. asymmetric, which means that if $a > b$ then it is never true that $b > a$,
3. transitive, which means that if $a > b$ and $b > c$, then $a > c$.

These three properties can easily be demonstrated by the technique used above. However, we must still ascertain whether the relation "greater" is independent of the particular form of the number couple. This means that if a number couple is greater than another, then this relation shall also be valid when the two number couples are replaced by anyone of their equals.

HYPOTHESIS. $(a, b) > (c, d)$
$$(a, b) = (a', b'), \qquad (c, d) = (c', d')$$

PROPOSITION. $(a', b') > (c', d')$

PROOF. The hypothesis says:
$$a + d > b + c$$
$$a + b' = a' + b$$
$$c + d' = c' + d$$

Let us add these three equations after we have interchanged the two sides of the middle equation. We obtain

$$a + b + c + d + a' + d' > a + b + c + d + b' + c'$$
or $$a' + d' > b' + c',$$

which is precisely the content of our proposition. Consequently we have shown that the relation "greater" has the three formal properties of the usual concept "greater", and furthermore that it is independent of the particular form in which the number couples are represented.

3. Definition of "smaller"

We say that a number couple is smaller than another, in symbols $(a, b) < (a', b')$, if $a + b' < a' + b$.

The very same considerations apply here as in 2. We now infer from the definition of „=", „>", „<" that the *number couples form an ordered system*. This means that between two number couples there always exists one and only one of the three relations "greater," "equal," "smaller." For, if we take any two number couples (a, b) and (a', b'), their order is established merely by comparing the two expressions $a + b'$

4. The Rigorous Construction of the Theory of Integers

and $a' + b$; these are natural numbers and as such it is already valid that they are related to one another by only one of the three relations; and this carries over to the number couples.

Furthermore, it is easy to prove that any positive number couple is greater than the null couple; the null couple is greater than any negative number couple; any positive number couple is greater than any negative number couple.

We will now show that one can operate with number couples in a manner wholly analogous to that with natural numbers. For this purpose we must define what is to be understood by the sum, difference and product of two number couples.

4. Definition of the sum
$$(a, b) + (c, d) = (a + c, b + d)$$

Before we accept this definition of sum we must test whether it has those properties which belong to the concept of sum in ordinary arithmetic. These properties are:

1. The sum always exists; this means that two numbers a and b always determine a third number, their sum.
2. The sum is uniquely determined.
3. The commutative law is valid: $a + b = b + a$.
4. The associative law is valid: $a + (b + c) = (a + b) + c$.
5. The monotony law is valid: if $b > c$, then $a + b > a + c$.

Only after we have shown that all these properties are satisfied, will we be justified[3] in calling the combining operation, introduced here, "addition."

As to 1. What does it mean to say that the sum exists? First, it means that the sum has again the form of a number couple, that is, when we combine two number couples additively, we remain in the domain of number couples; secondly,

[3] Cf. p. 63.

4. THE RIGOROUS CONSTRUCTION OF THE THEORY OF INTEGERS

that this number couple exists, that is, its components are again natural numbers. That both conditions are satisfied, follows immediately from the definition; for, since a, b, c, d are natural numbers, $a + c$, $b + d$ are also natural numbers.

As to 2. This condition means that the sum shall be independent of the particular form of the summands; in other words, if any number couple in a sum shall be replaced by a number couple equal to it, then the sum shall not be altered.

HYPOTHESIS: $(a, b) = (a', b')$, $(c, d) = (c', d')$
PROPOSITION: $(a, b) + (c, d) = (a', b') + (c', d')$
PROOF: If we perform the indicated additions, then we have to show that
$$(a + c, b + d) = (a' + c', b' + d');$$
this means
$$a + c + b' + d' = a' + c' + b + d.$$
This equation is an immediate consequence of the hypotheses if they are written as relations between natural numbers.

We will now drop this investigation to avoid being tiresome. In view of the above remarks, the reader should have no difficulty in proving the validity of conditions 3—5.

Before going on to subtraction we will draw attention to two consequences of our definition. First, we have that $(a, b) + (b, a) = (0, 0)$. Such number couples will be called "inverse" or "opposite" number couples.

Secondly, we have
$$(a, b) + (a, b) = (2a, 2b)$$
$$(a, b) + (a, b) + (a, b) = (3a, 3b)$$
and in general
$$\underbrace{(a, b) + (a, b) + \ldots\ldots + (a, b)}_{n \text{ times}} = (na, nb)$$

Instead of saying that the number couple (a, b) is set n times as a summand, we may also say that "we are forming the n-th

multiple of (a, b)"; for this sum we introduce the concise notation n.(a, b). Then we can express the result in the form:
$$n \cdot (a, b) = (na, nb).$$
This means that to multiply a number couple by a natural number, each term of the number couple must be multiplied by this natural number.

From these two results it now follows that *every number couple* (a, b) *can be represented in a definite normal form*, namely, in the form
$$a\,(1, 0) + b\,(0, 1).$$
For, if we perform the designated operations, we obtain:
$$a\,(1, 0) + b\,(0, 1) = (a, 0) + (0, b) = (a, b).$$
We usually call the number couples (1, 0) and (0, 1) the positive and negative units respectively, and introduce the notation
$$(1, 0) = +1$$
$$(0, 1) = -1;$$
furthermore, if we set:
$$a \cdot (1, 0) = a \cdot (+1) = +a$$
$$b \cdot (0, 1) = b \cdot (-1) = -b,$$
then we can represent every number couple in the normal form:
$$(a, b) = (+a) + (-b).$$

5. *Definition of the difference*

What shall we understand by the difference of two number couples? By analogy to the sum it is natural to set up the following definition:
$$(a, b) - (c, d) = (a - c, b - d).$$
However, this definition is open to criticism since, if c is greater than a or d is greater than b, then the differences a — c, respectively b — d do not exist. Hence in the domain of number couples subtraction could be performed only under certain conditions—on the other hand, the only reason for introducing

4. The Rigorous Construction of the Theory of Integers

number couples was to make subtraction into an operation which could be performed without restriction.

In order to avoid this difficulty we recall that the sum of a number couple and its inverse is the null couple. As we will see, the null couple occupies a position among the number couples which is very similar to that of zero among the natural numbers. Hence we can reduce the operation of subtracting a number couple to that of adding its inverse.

Accordingly we define:
(5) $\quad (a, b) - (c, d) = (a, b) + (d, c) = (a + d, b + c)$.
This definition is free of the defect which disturbed us earlier. The number couple standing on the right always exists; consequently, subtraction can be performed without restriction.

By the above remarks it follows that the operation defined in (5) actually has the properties of subtraction. First of all it is the inverse of addition. It is uniquely determined and satisfies neither the commutative nor the associative law.

Before leaving the consideration of subtraction we will give an application of definition (5). From (5) it follows in particular that
$$(a, 0) - (b, 0) = (a, b),$$
a result which we can also express in the form
$$(a, b) = (+ a) - (+ b).$$
Hence every number couple can be represented as the difference of two positive numbers.

If we take into consideration the representation derived on p 36, it follows that
$$(+ a) - (+ b) = (+ a) + (- b).$$
Hence we could say: if we write, for the sake of brevity, $(a, b) = a - b$, then it follows that a number couple is simply the difference of two ordinary numbers. However, we resist this temptation and insist on the more complicated way of writing $(a, b) = (+ a) - (+ b)$ for definite reasons.

Finally, it must be realized that the plus and minus signs appearing in the last expression have entirely different functions. The plus signs are signs serving as *prefixes*, while the minus sign is a sign indicating an *operation*. The plus sign in + a is actually only an abbreviation for (a, 0) and can be dispensed with if we write the number couples themselves. This is not true, however, of the minus sign which combines the two number expressions.

6. *Multiplication*

From the interpretation of sum it followed that n (a, b) = (na, nb). Hence we already know how to form a multiple of a number couple. However, this does not tell us how one number couple is to be multiplied by another. Our previous analysis does not give us a starting point for this operation. On the contrary, we must set up a new definition, and so we define

(6)   (a, b) · (c, d) = (ac + bd, ad + bc).

This is an arbitrary convention. It is only justified by the fact that this operation actually has some especially important formal properties of multiplication. For this purpose we have to prove the validity of the following propositions:

1. The product of two number couples is always another number couple.
2. The product is uniquely determined, independent of the particular form of the number couples.
3. It satisfies the commutative law.
4. It satisfies the associative law.

We will suppress the detailed proofs here as they will probably be too tedious for the reader. Let us suppose therefore that the proofs have been carried through. Then the product of number couples agrees with the product operation of ordinary arithmetic as far as the above properties are concerned; this gives us the right to bestow on this operation the name "multiplication." However, all properties of ordinary

4. THE RIGOROUS CONSTRUCTION OF THE THEORY OF INTEGERS

multiplication do not carry over to the new operation. For example, the "monotony law" is no longer valid. This law states that $B > C$ implies $AB > AC$; it is valid here if and only if the factor A is positive.

Our definition (6) yields, in particular, the following relations:

$$(1, 0) \cdot (1, 0) = (1, 0)$$
$$(1, 0) \cdot (0, 1) = (0, 1)$$
$$(0, 1) \cdot (1, 0) = (0, 1)$$
$$(0, 1) \cdot (0, 1) = (1, 0)$$

These are merely the rules of signs:

$$(+1) \cdot (+1) = +1$$
$$(+1) \cdot (-1) = -1$$
$$(-1) \cdot (+1) = -1$$
$$(-1) \cdot (-1) = +1$$

Hence the rules of signs follow from (6); this definition of multiplication is in itself arbitrary and was laid down in the specified manner for the express purpose of rendering the most important properties of the multiplication of natural numbers.

This fact was already known to Gauss. In the year 1811 he wrote to Bessel: "We should never forget that the functions, as all mathematical combinations of concepts, are only our own creations, and that when the definition, from which one proceeds, ceases to have a sense, one should not ask, strictly speaking, *what has to be assumed?* but *what is convenient to assume?* so that I can always remain consistent. Thus, for example, the product of minus by minus." And Hankel, in the treatise already quoted, has this to say about the rules of signs: "It cannot be stressed strongly enough that, in spite of a widespread general opinion, these equations can never be proved in the formal arithmetic; they are *arbitrary conventions* for the purpose of preserving the formalism in the calculus."

7. *Division*

A remark must finally be made regarding division. We already know that this operation takes us out of the domain of integers. How does this fact manifest itself in the calculus that we have constructed?

For the time being, let us express the result of the division in terms of unknowns:
$$(a, b) \div (c, d) = (x, y).$$
If this division is to be the inverse of multiplication, then
$$(a, b) = (c, d) \cdot (x, y)$$
which means: $(a, b) = (cx + dy, dx + cy)$;
this equation implies according to definition (1) that
$$a + dx + cy = b + cx + dy.$$
Consequently, the division can be performed only if two natural numbers x, y can be found which satisfy this last equation. Now this equation is an example of a "Diophantine equation," and from number-theoretic investigations (into which we will not go here) we know that such an equation has a solution only under very definite conditions. (For instance, the equation $3x = 2y + 4$ certainly does not have a solution among the natural numbers, since the right side of the equation is divisible by 2 while the left side is not.) Hence division cannot be performed unlimitedly in the domain of integers.

Let us now look back over the course we have pursued. Our problem was: how can the integers be based on the system of natural numbers alone without using either a hypothesis or a postulate? The submitted construction gives the answer to this question. We have erected new concepts out of the natural numbers and, by virtue of arbitrary but appropriate conventions, endowed them with such properties as will enable them to perform everything integers are supposed to perform. Now we only have to take the last step and address outright as numbers the concepts we have created.

4. THE RIGOROUS CONSTRUCTION OF THE THEORY OF INTEGERS

Only one point remains somewhat problematical. Can't our conventions be in conflict with one another? As long as we perceived the facts of the arithmetic of integers as resting, let us say, on intuitive insight, the question was meaningless; however, in a system whose basic laws have the character of arbitrary conventions, no guarantee exists that these will not lead to a contradiction. A formal theory demands the proof of consistency, and this proof will form the keystone of the entire structure.

In formulating the operational rules for number couples we used the calculus of differences as a model; our six definitions are exact copies of the six formulae on p. 27. If a contradiction could be derived from our conventions, the same chain of reasoning must produce a contradiction in the domain of cardinal numbers, where it would refer to the corresponding formulae in the calculus of differences. This follows from the fact that the formulae have the very same structure, and the process of drawing conclusions depends only on this structure. Accordingly, these formulae would also have to contain a contradiction in themselves. Consequently we recognize that the arithmetic of number couples is certainly free of contradictions if this is true of the arithmetic of natural numbers.

Before closing we will look into an important question. How are the integers related to the natural numbers? Are they an extension of the latter? For a long time this was the opinion of mathematicians. In this interpretation the natural numbers are set equal to the positives and are then called "positive numbers" merely to distinguish them from the negative numbers. However, even colloquial usage suggests that it is well to distinguish between the counting and the positive numbers. If I say that I have invited three guests, the numeral "three" can here be replaced by any other numeral. However, it cannot be replaced by the symbol "$+3$," since otherwise it would also have a sense to speak of inviting -3 guests. "3" and "$+3$"

4. The Rigorous Construction of the Theory of Integers

have, so to speak, a different logical grammar; the carefully erected structure of arithmetic confirms this. The negative numbers are not a later complement of the natural numbers. In reality they are two entirely separate number systems. This stands out very clearly by our construction, for the very fact that the positive numbers are number couples, sharply distinguishes them from the natural numbers. This is why we have not interpreted positive numbers as differences of two ordinary numbers.

Hence natural numbers and integers form two distinct systems; they are, so to speak, on two different levels. However, there exist certain relations between them which under superficial examination give the impression that the first system is a part of the second. We will look into the nature of these relations. To every natural number there corresponds a positive number

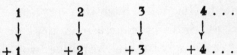

such that the order and combination of the natural numbers is exactly the same as the order and combination of the corresponding positive numbers. To speak more precisely, the correspondence between the two sequences has the following three properties:

1. It is one-to-one, that is, every natural number corresponds to a positive one and conversely.

2. The natural ordering of the individual numbers is preserved by this correspondence, that is, if a is greater (smaller) than b, then the corresponding positive number $+\,a$ is also greater (smaller) than the corresponding positive number $+\,b$.

3. The combination of the numbers in the one sequence by the four basic rules is mapped on the same combination of the corresponding numbers in the other sequence. For example, if $a + b = c$, then it is also true that $(+\,a) + (+\,b) = +\,c$; and analogously for the other operations.

4. The Rigorous Construction of the Theory of Integers

To describe this concisely, we will say that the correspondence is *one-to-one, similar* and *isomorphic*. This is the basis for the fact that all relations between natural numbers are valid for positive numbers; thereby one system appears as a true copy of the other.

In general, if two number systems are submitted, S with the elements a, b, c, ... and S′ with the elements α, β, γ, ..., a mapping between S and S′ is said to be an isomorphism if the result of the combinations $\alpha + \beta, \alpha - \beta, \alpha \cdot \beta, \alpha : \beta$ always corresponds to the result of the combinations $a + b, a - b, a \cdot b, a \div b$.

We should not assume that a relation which is one-to-one and similar is necessarily isomorphic. The following is a simple counter-example. Let us associate the sequence of natural numbers to the sequence of even numbers as indicated by the arrows:

```
1    2    3    4    5    6  ...
↓    ↓    ↓    ↓    ↓    ↓
2    4    6    8   10   12  ...
```

The correspondence thereby determined is single-valued and similar, but not isomorphic. For example, the product of the numbers 2 and 3 in the first row is 6, the product of the corresponding numbers in the second row is 24; however, 6 and 24 do not correspond to one another.

To recapitulate: The natural numbers are not a subset of the integers. However, the latter contain a proper subsystem—the positive numbers—which can be mapped on the system of natural numbers by a correspondence which is one-to-one, similar and isomorphic.

Now what can the new system do? Can it be used to solve problems which could not be solved with the old system? In regard to this question, too, we will come to a somewhat different understanding.

Strictly speaking, the problem, to take the number 7 away from the number 4, is not solved by the introduction of nega-

43

tive numbers; rather another problem is solved, namely, to take the number $+7$ away from the number $+4$. This new problem is the analogue of the original one in the domain of integers. Hence an unsolvable problem never changes into a solvable one by the invention of new numbers—such an interpretation would be somewhat superficial and false—instead the problem projects itself, so to speak, into another number level and there the *corresponding* problem has a solution.

Historically, of course, things did not develop in this way. The negative numbers were not suddenly introduced into mathematics; instead they forced themselves, as it were, into the practice of counting. They are probably an invention of the Hindus, who are also responsible for our system of numerals. In the occident they first appeared at the time of the Renaissance, just as the operating with letters had become accepted. Above all, it was the problem of solving equations which, with a certain inner necessity, led to the negative numbers. For example, Chuquet in his treatise *"Le Triparty en la Science des Nombres"* (1484) arrived at negative solutions when he considered the problem of decomposing a number by a fixed rule. At first these new numbers were regarded with some distrust. This is attested by designations prevalent in those days such as *absurde Zahlen* (absurd numbers) (M. Stifel), *numeri ficti* (fictitious numbers) (G. Cardano) etc. An algebraist as prominent as Vieta (1540 to 1603) would exclude them on principle from mathematics—an attitude taken by many English mathematicians up to the end of the 18th century, whose objections were aimed mainly at the shady foundation of the rule of signs. Even prominent mathematicians did not have a very clear picture of these relations; for example, Wallis (1616 to 1703) perceived the negative numbers as "supra-infinite" quantities.

He argued as follows. The numbers

$$\frac{1}{3}, \frac{1}{2}, \frac{1}{1}, \frac{1}{0}$$

form an increasing sequence; in general

4. The Rigorous Construction of the Theory of Integers

$$\frac{1}{m} < \frac{1}{m-1},$$

for $m = 0$ this gives $\quad \frac{1}{0} < \frac{1}{-1}$,

which means: $\infty < -1$. Hence the negative numbers are greater than ∞, "plus quam infiniti."

In reality the situation is as follows: $\frac{1}{0}$ is not a symbol to which a relation of magnitude can be ascribed; hence the formula
$$\frac{1}{m} < \frac{1}{m-1}$$
is not valid if $m = 0$. A clear insight into these matters was first attained in the 19th century; in M. Ohm's *Versuch eines vollständig konsequenten Systems der Mathematik* (Attempt at a Perfectly Consistent System of Mathematics) (1822) the principle of the extension of number domains was explicitly stated for the first time. The leading position of the permanence principle was then recognized by H. Hankel.

We could give, by the way, a clear meaning to the idea of Wallis if we altered slightly our conception of the number *axis*. For this purpose we will construct another geometrical model of the real numbers, the "number circle." If we map the points of the number axis by "central projection" on the points of a

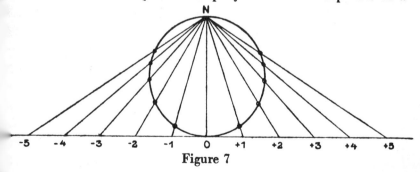

Figure 7

circle, as shown in the accompanying figure, it is clear that every positive and negative number corresponds to a definite point on the number circle and conversely. Only one point is an exception, for it is obvious that the center of projection N does not correspond to a point on the number axis, in other

4. THE RIGOROUS CONSTRUCTION OF THE THEORY OF INTEGERS

words, to a real number. Hence we would have to say that between the points of the circle and the real numbers there exists a one-to-one correspondence with the single exception of the point N. Now the mathematician strives, whenever he meets exceptions, to find a formulation which will remove the exception. Thus, the theorem "every quadratic equation has two roots" can be maintained only if we decide, in the cases where two distinct roots do not exist, to count the one root twice. So we will now introduce a real number which corresponds to the point N. If a point moves further and further out on the number *axis* (it is all the same whether it moves to the left or to the right), its image advances nearer and nearer toward N. We will now say that the straight line has a *single infinitely distant* point, the image of the number ∞.

The reader will object to this assumption. He will say that the points of a straight line are objectively given and we are not free to insert a new point on it. We are certainly not allowed to dispose of the points of a line, but we may very well dispose of our way of speaking. And the one proposed here has, at any rate, the merit of closing any gap in the correspondence between the points on the circle and the real numbers. On receiving the "improper point" the line becomes the bearer of a "cyclic order." We can now attribute the following significance to Wallis' idea. Usually we go from the positive to the negative numbers by passing through 0 however, we could also take a path over ∞, and in this sense Wallis is entirely correct when he denotes the negative numbers as supra-infinite. "What is the case in reality? Are the negative numbers greater or smaller than the positives?" The reader will note that these two interpretations amount to two different orderings

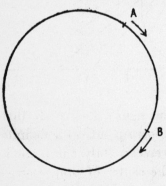

Figure 8

4. The Rigorous Construction of the Theory of Integers

of the real numbers. Let us put it this way: if we include the improper number ∞ among the points of the number axis the number axis is thereby closed just as the circle. In the case of the circle we cannot say, for instance, which point precedes the other. For example, we can go either from A to B or from B to A, indeed moving in the same direction. Furthermore, if we select three arbitrary points on the circumference, we can justly assert that each lies between the other two. Hence the concept of "betweenness" loses its meaning. In short, we are concerned here with an order which is totally different from that on a straight line.

However, we can also impress a linear order on the points of a circle. For this purpose we only have to delete one point, that is, to break the connectedness of the line, and to run from this gap through the points in a fixed, though arbitrary, direction. If we delete the point ∞ we obtain the usual ordering of the real numbers (Fig. a); however, if we delete 0, we obtain the ordering of Wallis (Fig. b).

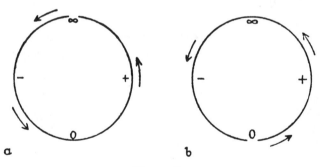

Figure 9

This example shows us that it is not always possible to extend the number domain in only one way. Actually, we have just obtained two entirely distinct number systems. Why does this situation arise? The underlying reason in this case is merely that the *permanence principle*, which should guide us in the extension of the number domains, falls short, or, more exactly, that this principle no longer uniquely regulates the formation of concepts. It only requires that in extending the system of concepts certain rules, especially important to us, remain valid, not that all remain intact. Actually we must resign ourselves to the fact that in the ascent to a higher number domain the validity of certain calculating rules of the old domain is lost. For instance, the step from the natural numbers to the integers is taken only at the price that we renounce the validity of the monotony law of multiplication. Hence, in general, only a part of the old calculating rules will carry over to the new higher domain. It can now happen that various paths are available for the extension of the number domain and that a different calculating rule must be renounced in each case. This is exactly the case here. Thus, either we hold fast to the validity of the formula

$$m > m - 1$$

for $m = 0$, and then we must sacrifice the formula

$$\frac{1}{m} < \frac{1}{m-1}$$

Or we can proceed with the extension so that the second formula remains intact, and then we must abandon the first. To recapitulate: The permanence principle is not a dependable guide. It allows us the choice of two paths which lead to distinct orderings of the real numbers and therefore to two distinct number systems.

5. The Rational Numbers

The rational numbers are much older than the negative numbers. While the latter did not appear until a certain stage of development of mathematics was reached, namely, the stage of algebra, the introduction of fractional numbers had already been forced by the needs of daily life, perhaps by the development of a system of weights and measures. This can also be recognized in the use of language. For, while a notation for positive and negative numbers appeared only in the seventeenth century, even the colloquial language of civilized peoples already had expressions for fractional parts. However, the concept of rational number, as we know it today, crystallized only at a comparatively late date. "From our modern point of view the various fractional concepts are no longer essentially distinct from one another. We think of $1/2$, $1/20$, $17/25$ as numbers of the same type; our uniform notation for fractions in terms of numerator and denominator also stresses this fact. However, this was not always the case. For example, the Egyptians did not have a symbol at all for fractions such as $7/25$; for $1/20$ they used a symbol which did not take into account the numerator 1; and for $1/2$, a symbol which was quite different from that used for the other fractions." O. Neugebauer, *Vorlesungen über Geschichte der antiken mathematischen Wissenschaften* [Lectures on the History of Antique Mathematical Sciences], Vol. 1, p. 86.) It seems that in earlier times certain fractions such as $1/2$, $1/3$, $2/3$ were looked upon as independent and individual numbers, which

49

5. THE RATIONAL NUMBERS

are not derived as secondary formations from the corresponding integers. These "natural fractions," which might, perhaps, have a more metaphorical qualitative meaning, were incorporated into a system only later in the development of arithmetic. It was only through this process that there arose what is called today the system of rational numbers.

At any rate, from an historical point of view, we face processes which are complicated because, to all appearances, the thread of their development is taken up from the means of expression already in existence. Processes are never indifferent to the power of a definite system of symbols. The existing system of writing determines in a wide sense the potential development of mathematics. Just as the restriction to a definite building material (freestone, brick, wood) pushes architecture toward certain forms—produces building styles—so also does the use of a definite notation generate something like a style of mathematical thought whenever the notation contains certain expressions but not others. For example, the Egyptians had a notation by which they could only express the reciprocal of an integer and its complement (namely, the fractions $\frac{1}{n}$ and $1 - \frac{1}{n}$). As a result, Egyptian mathematics was pressed into very definite types of questions (decomposition of unit fractions), while its approach to other problems was as good as prohibited.

Consequently, a definite style arose in stating and answering questions. The Greeks, who could express mathematical concepts only by means of geometrical figures, failed to note certain possibilities of modern mathematics and, above all, to comprehend the general concept of number. On the other hand, the fact that the Babylonians developed an algebra of astonishingly high order seems closely connected with the structure of their speech and writing.

For the moment we wish to try to formulate somewhat more clearly the questions which force themselves upon u

5. The Rational Numbers

regarding these matters. We use today two kinds of writing, for a symbol represents either a sound, as "r" or "o", or a concept, as "3" or "+". Accordingly, we distinguish between phonetic writing and symbolic writing (ideography). All symbolic writing, as we know today, originally arose from a pictorial writing which reproduced only concrete perceptible things or events. The question now arises: out of an intuitive pictorial writing, how could an abstract mathematical formalism detach itself with conventional special symbols for quantities and operations? Or, out of the letter writing which followed the phonetic picture of the words, how could the concept symbols of mathematics spring up? Both are difficult to understand, since the mathematical symbols do not lie, so to speak, in the direction of the natural development of language. And still the presence of ideographic writing is essential for the unfolding of mathematical thoughts. In Egypt, with its historical continuity, this step did not take place. In Babylon, where two entirely distinct cultures, the Sumerian and the Akkadian (with languages of basically distinct grammatical types) are superimposed, the path for such a formal development was cleared. Namely, through the contact of these two distinct languages there arose the possibility of writing a word either by syllables or by ideograms. In the Akkadian texts both modes of writing are arbitrarily used in turn; thereby there arose the possibility of writing the mathematical concepts (quantities and operations) ideographically and of attaining a language of formulae, while the remaining text is written in syllables. It was therefore not rational considerations, but rather historical chance—the meeting of different cultures—which led to the formation of a language of mathematical symbols. Perhaps it is "that in the frame of a continuous historical development, which actually depends on the traditions handed down from generation to generation, the awareness of the arbitrary and purely conventional character of all modes of expression does not arise at all; instead, all these things exist as forms,

5. The Rational Numbers

absolute and given, which to change essentially and voluntarily greatly exceeds the analytic faculty of men. Only men who descend from an entirely different historical tradition are freely able to use the foreign means of expression and to recognize their limits as well as their possibilities." (Neugebauer, p. 78.)

We have referred in all these matters to the lectures of Neugebauer, who investigated the interrelation between mathematics and language. We now wish to direct *our* inquiries in another direction. Accordingly, we will not investigate how the operational rules of fractions developed and what factors have started them on their way. Instead we will be interested only in the question *by what right* are the fractional numbers introduced in arithmetic. Just as we constructed the integers out of the natural numbers, so also will we now construct the rational numbers with the help of the integers. Since the basic propositions of this technique have already been discussed in the previous chapter, we will now be able to do this much more concisely.

To prepare the way for further construction we will consider the following questions. Let us suppose that we are dealing with divisions which can be carried out without a remainder. Can we operate with the results of such divisions without actually carrying them out? It is rather natural to think of the fact that the divisions 48:3 and 80:5, for example, produce the same result, and this fact must make itself known somehow by the numbers appearing in the calculation. Also, by merely looking at two divisions, could we not tell which would produce the greater result? In short, could not a calculus of quotients be set up in which we actually operate only with dividend and divisor? We will immediately see that this is possible. In doing so we will remain throughout within the domain of integers. This means that every proposition to be considered will be demonstrable within the arithmetic of integers. In order to obtain the fractions we will drop the

5. THE RATIONAL NUMBERS

assumption that the division shall be carried out without a remainder, and we shall speak of the propositions which are valid in the arithmetic of integers as the *operational rules for arbitrary number couples*. The calculus of number couples will then tie itself in a very natural way to the calculus of quotients. In accordance with our intentions we now ask the questions:

1. When are two quotients equal?
2. When is a quotient greater or smaller than another?
3. How can quotients be added, subtracted, etc.?

As to 1. What relation must exist between four numbers a, b, c, d, if the divisions a:b and c:d are to yield the same result?

We may say that in this case the numbers a, b, c, d must form a proportion, and it is known that this is possible only if $ad = bc$. But, the theory of these proportions, as it is usually represented, already assumes the system of rational numbers, which we have not yet constructed. Hence we must proceed with caution.

First, we prove that the result of a division remains unchanged if the dividend and divisor are multiplied by the same number; that is, if

$$a : b = q$$

then it shall also be true that

$$n\,a : n\,b = q$$

Now the first equation implies

$$a = b \cdot q$$

On multiplying both sides by n,

$$n\,a = n\,b \cdot q$$

This is the meaning of the second equation. (The reader should note that in this derivation we have never stepped out of the domain of integers.) In the very same way we can show that two quotients with equal divisors are equal only if their dividends are also equal.

53

5. THE RATIONAL NUMBERS

The proposition, which we have just proved, can be used to determine whether any two divisions
$$a : b \text{ and } c : d$$
yield the same result. For this purpose the two quotients are first transformed so that they have the same divisors. Then they will yield the same result only if their dividends are also equal. This means that if the terms of the first quotient are multiplied by d, the second by b, so that the quotients
$$ad : bd \text{ and } bc : bd$$
are obtained, then equality will follow only if
$$ad = bc.$$

As to 2. The above remarks also enable us to answer the second question. The division $a : b$ yields a greater, respectively, smaller result than the division $c : d$ according as ad is greater, or smaller, respectively, than bc.

Summarizing we can say:

$$a : b \gtreqless c : d \text{ according as } ad \gtreqless bc$$

As to 3. How can a division be formed whose result is exactly as large as the results of two other divisions added together? The division
$$(ad + bc) : bd$$
has the required property. First
$$(ad + bc) : bd = (ad : bd) + (bc : bd)$$
is valid by a proposition which can be proved in the arithmetic of integers. Secondly, a common factor can be cancelled from each of the two divisions on the right. We thereby obtain
$$(ad + bc) : bd = (a : b) + (c : d).$$

In an entirely analogous manner the difference of quotients can be formed.

The product of quotients is obtained from the product of the dividends and the divisors. Thus, if
$$a : b = q \text{ and } c : d = r,$$
then
$$a = bq \text{ and } c = dr$$

5. THE RATIONAL NUMBERS

By multiplication we obtain
$$ac = bd \cdot qr, \text{ which means } ac : bd = qr.$$
This implies that the last division yields a result which is exactly as large as the product of the results of the two original divisions. It is left to the reader to find a proof that
$$(a : b) : (c : d) = ad : bc.$$
What have we gained by these propositions? The insight that we can calculate with quotients just as with numbers. We come now to the critical step whereby we will extend this calculus to arbitrary number couples and let the formulae proved above serve as definitions of the concepts "equal," "greater," "smaller," "sum," "difference" etc.

1. *Definition of equality*

For the time being we think of number couples as entities which have as yet no other meaning associated to them than that of being number couples. We define:
$$(a, b) = (c, d) \text{ if } ad - bc = 0.$$
From the previous chapter we know the conditions which must be satisfied by an arbitrary definition of the concept of equality, namely, the relation so defined must be reflexive, symmetric, transitive. It is obvious that the first two properties are satisfied.

On the other hand, the proof of the transitivity requires a little reflection. We have to show that
$$(a, b) = (a', b')$$
$$\text{and } (a', b') = (a'', b'')$$
always implies
$$(a, b) = (a'', b'').$$
By (1) the first two equations state that
$$ab' - a'b = 0$$
$$a'b'' - a''b' = 0,$$
and the conclusion to be drawn is that
$$ab'' - a''b = 0.$$

55

5. THE RATIONAL NUMBERS

If we multiply the first equation by b″, the second by b and add both, we obtain
$$ab'b'' - a''bb' = 0$$
which means $b'(ab'' - a''b) = 0$

Does this mean that $ab'' - a''b = 0$? Only if b′ is different from 0. If it so happened that $b' = 0$, then the first number couple could very well be equal to the second, the second equal to the third, without the first being equal to the third. The transitivity is therefore guaranteed only if the case $b' = 0$ cannot arise. Accordingly, we require that those number couples shall be excluded whose second term is 0 (in the usual terminology, this means that we exclude the division by 0).

A simple consequence of this definition is that every number couple can be put into infinitely many forms. Thus
$$(a, b) = (2a, 2b) = (3a, 3b) = \ldots$$
(This says that the numerator and denominator of a fraction can be multiplied by an arbitrary number without changing the value of the fraction.)

2. *Greater and smaller*

We define

$(a, b) < (c, d)$, if $ad - bc < 0$
$(a, b) > (c, d)$, if $ad - bc > 0$.

The proof, that the relation so defined is irreflexive, asymmetric, transitive, presents no difficulty. We pass over it.

From definitions (1) and (2) the important conclusion can now be drawn that number couples form an *ordered system*, in other words, two number couples are always related to one another by one of the three relations "greater," "equal," "smaller." This follows immediately as soon as we bear in mind that the expression $ad - bc$ is an integer, and therefore it must certainly be either positive or zero or negative; obviously this disjunction carries over to the number couples.

3. *Addition*

The sum of the two number couples (a, b) and (c, d) is defined as the number couple (ad + bc, bd). Before we adopt this definition we must test whether it satisfies the formal requirements which we must place on the concept of sum.[1]

a) The sum shall *exist*. This implies two conditions: first, the sum shall have the form of a number couple; secondly, each of its two terms shall be an integer. Not every definition would satisfy these conditions. Thus, if we were to define, let us say,
$$(a, b) + (c, d) = (a, b, c, d),$$
the sum would no longer belong to the domain of number couples. Furthermore, definitions are conceivable by which the first or the second term become fractional numbers. Our definition avoids both obstacles.

b) The sum shall be uniquely determined by the summands. We have to attribute some importance to this condition because a number couple has infinitely many distinct representations, and we are not permitted to assume beforehand that the sum is independent of the form of the number couples. In other words, if we form the sum
$$(a, b) + (c, d) = (ad + bc, bd)$$
and replace (a, b) and (c, d) by number couples (a', b') and (c', d') equal to them, then it shall also be true that
$$(a, b) + (c, d) = (a', b') + (c', d').$$
We can easily verify by a calculation that this condition is satisfied.

c) The addition is furthermore *commutative* and *associative*, and finally we have

d) the *monotony law*, which says that $A > B$ always implies $A + C > B + C$.

---

[1] The reader whose main interests are philosophical may omit the further development to p. 61.

5. The Rational Numbers

From the definition of sum it follows in particular that
$$(a, b) + (a, b) = (2a, b),$$
a result which can be concisely written as
$$2(a, b) = (2a, b);$$
in general, we have
$$n(a, b) = (na, b),$$
which tells us how to multiply a number couple by an integer.

4. *Multiplication*

We define
$$(a, b) \cdot (c, d) = (ac, bd).$$

This definition must satisfy conditions analogous to those stated for the sum. We omit the proof that conditions (a) to (c) are satisfied and only remark that condition (d), the monotony law of multiplication, can no longer be maintained.

Moreover, addition and multiplication satisfy the distributive law
$$A(B + C) = AB + AC.$$

5. What is the status of subtraction and division? Must these be defined anew? This is not necessary, since they are completely determined as soon as we interpret them as the inverse operations of addition and multiplication. Thus, let us assume that
$$(a, b) - (c, d) = (x, y),$$
implies the statement
$$(a, b) = (c, d) + (x, y);$$
by definition (3) we have
$$(a, b) = (cy + dx, dy),$$
and by (1) this equation can only exist if
$$ady - b(cy + dx) = 0,$$
namely, if
$$(ad - bc)y - bdx = 0.$$

5. THE RATIONAL NUMBERS

At any rate, a solution of this equation is given by
$$x = ad - bc$$
$$y = bd;$$
the other solutions are obtained by multiplying this solution by arbitrary integers. But, by the remark on p. 56, the number couple (nx, ny) is equal to the number couple (x, y). Hence we can always represent the result of a subtraction in the form
$$(a, b) - (c, d) = (ad - bc, bd).$$
We can deal with division in an entirely similar manner. If we set
$$(a, b) : (c, d) = (x, y),$$
then
$$(a, b) = (c, d) \cdot (x, y),$$
which implies
$$(a, b) = (cx, dy)$$
or
$$ady - bcx = 0.$$

The solution of this equation is (if we again disregard the integral multiples)
$$x = ad, \, y = bc.$$
Hence we obtain
$$(a, b) : (c, d) = (ad, bc).$$
If we call (d, c) the reciprocal of (c, d), then this result can be expressed as follows: to divide a number couple by another, multiply it by the reciprocal of the latter.

The conceptual objects, which we have constructed, are therefore ordered and we can calculate with them just as if they were natural numbers. These two facts induce us to designate the new concepts as numbers, to be more explicit, as *rational numbers*. In the system of rational numbers we have created an instrument with which the four basic rules of arithmetic can be performed unlimitedly, with the exception of the division by zero.

We will also think of the consistency of the new system as based on the fact that the definitions of the concepts

5. THE RATIONAL NUMBERS

"greater," "equal," "smaller" as well as the operational rules are exact copies of propositions in the calculus of quotients, so that every contradiction in the system of rational numbers must also become one in the system of integers.

Among the integers the numbers 0 and 1 play a special role. For every number a we have
$$a + 0 = a, \quad a \cdot 1 = a.$$
This means that a number is reproduced if it is combined either with the number 0 as a summand or with the number 1 as a factor. We say that 0 is the modulus of addition and 1 the modulus of multiplication.

Among the number couples the couples $(0, 1)$ and $(1, 1)$ play an entirely analogous role; this means that
$$(a, b) + (0, 1) = (a, b)$$
$$(a, b) \cdot (1, 1) = (a, b)$$
On this account we will call these number couples the "rational number 0" and the "rational number 1." The facts demonstrated for the integers 0 and 1 are valid in general, since to every (positive and negative) integer a there corresponds a rational number $(a, 1)$. For example, we have
$$(a, 1) + (b, 1) = (a + b, 1)$$
$$(a, 1) \cdot (b, 1) = (a \cdot b, 1).$$
Consequently, the system of integers can be mapped on a part of the rational numbers, so that the ordering of the numbers as well as the combinations by the four arithmetical operations remain unchanged. More clearly stated, this means that the system of rational numbers contains a subsystem which can be mapped on the system of integers by a one-to-one, similar and isomorphic correspondence. Hence the rational numbers were thought of as an extension of the integers, including, besides the integers, also the fractions. This was a mistake. Actually each of the three systems—the natural numbers, the integers, the rational numbers—forms a closed system by itself, and it is quite impossible to go from one of these domains to another by adjoining new elements.

5. The Rational Numbers

We must therefore carefully distinguish between the natural number 5, the integer +5, and the rational number $\frac{5}{1}$. These three numbers are not identical, since they are elements of distinct calculi. They only correspond to one another, which means that they play a different role in their calculi.

This discussion prompts us to ask: do the arithmetical operations have the same meaning in each of these calculi? For instance, in the domain of integers, is subtraction the same operation as in the domain of natural numbers? Furthermore, what do we mean here by "the same?" If this implies that the operations must satisfy the same conditions, then the questions must be answered negatively. For, in the domain of natural numbers the expression $a - b$ is admissible only if $a > b$, while in the domain of integers this restriction is removed; obviously this is an important distinction. Consequently, there is not, strictly speaking, one subtraction but as many different operations with this name as there are domains of numbers. We should not be deceived regarding this situation by the fact that we use the same signs $+$, $-$, $:$, etc., at the various levels. If we put the statements of these concepts side by side, it becomes clear how far the analogy between them goes and where it stops.

We can continue these considerations still further. If an operation in a new domain does not satisfy all requirements that we are accustomed to place on it in the old, should we still call it the same operation? As an example, let us consider the sum of two rotations of a sphere. Everyone understands this to be the result of performing one rotation after the other; as such the terminology seems natural. However, does this concept of sum satisfy our five formal conditions?

a) The sum exists, since two rotations, performed one after another, always yield another rotation.

b) The sum is uniquely determined by the two rotations.

5. THE RATIONAL NUMBERS

c) Does the sum obey the commutative law? In other words, is the result of two rotations independent of the order in which these rotations are performed? We wish to see if this is true or not. For this purpose take two rotations of the globe through 90°: one around the NS axis of the earth, so that P goes into Q; the other, around the horizontal axis PP', so that N goes down to Q. What happens if we perform these two rotations one after the other? If we rotate first around the vertical axis, secondly around the horizontal, N goes to Q; if we perform the rotations in the reverse order, N goes over to P'. These two positions of the sphere are entirely different; therefore, the composition of two rotations is no longer commutative.

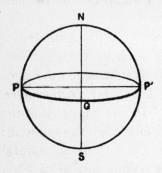

Figure 10

d) On the contrary, it can be proved that the composition is associative.

e) The fifth condition, the validity of the monotony law, is omitted, since we have not defined when a rotation of the sphere is greater or smaller than another.

In our example only three conditions are satisfied of the five that are enumerated above. It is up to us whether we still wish to use the word sum in this case.

Furthermore, we call the resultant of two forces their sum. In this case, conditions a) to d) but not e) are valid, since forces are not linearly ordered.

As further examples we may cite the addition of two waves or the sum of two colors (the "sum" of two colors could be, for instance, the color resulting from mixing the two colors). Hence by analogy the concept of sum may be applied to various realms of thought; however, we must realize that the word "sum" does not have the same connotation in these various applications.

5. The Rational Numbers

These remarks throw a light on many favorite expressions of philosophers, such as "The whole is more than the sum of its parts"; "The melody is more than the sum of its notes," etc. Such a statement says nothing, it is by no means clear what is to be understood by the sum of notes; this expression has not been defined as yet. Of course a meaning can be given to this phrase and then, perhaps, some correctness can be visualized in that assertion. However, before the correctness or non-correctness of the assertions under consideration can be discussed at all, we must first state *what* is to be understood by a "sum of notes."[2]

Considerations such as the above, can also be applied to the concept of equality. We have said that a relation shall be called an equality only if it is reflexive, symmetric and transitive. However, is the adherence to these conditions actually indispensable? On some further reflection we will come to another view. For example, it is not absolutely necessary that the concept of equality be transitive. In the comparison of perceptually given segments within one's field of vision, it frequently happens that the segments a and b appear equal in length, similarly for b and c, while a and c appear to have different lengths. Analogous remarks are valid regarding the agreement of waves, musical pitch, etc. The best way to account for this state of affairs is to say that the concept of equality, applied to such phenomena, is certainly symmetric but not necessarily transitive.

These considerations break down the preconceived opinion that the concept of equality *must* have certain properties. Instead we will say that concepts can be formed which are more or less related to the concept of equality between natural numbers, and it would only be inappropriate to call concepts with entirely different properties by the same name. Accord-

---

[2] Cf. Schlick, *Über den Begriff der Ganzheit*, Erkenntnis (*On the Concept of Wholeness*, Knowledge), Vol. 5.

5. The Rational Numbers

ingly we will improve the statement on p 29; instead of saying that the procedure described there is "justified," we will say that it is "appropriate." More precisely stated, this means that the conditions which we impose on the concept of equality for number couples will be divided into two parts:

a) *appropriateness* of the equality definition for number couples; thereby we demand permanence of the properties of the equality relation.

b) *justification* in the sense that for those couples which shall correspond afterwards to particular numbers, the same relation shall exist as for the particular numbers.

Before closing we wish to make a few more remarks.

1. If someone asks, what is a rational number, the best answer that can be given is to describe the calculus of these numbers, that is, to state the rules by which we calculate with such numbers. We often imagine that somehow the rational numbers "are there" already, independent of the calculus, and that the rules of calculation *follow* from the nature or the essence of these numbers. However, this is not the case, and our discussions show that the *concept* of rational number is determined through the *calculus* of the rational numbers.

The experience that we are able to divide segments, weights, etc., is not the basis of the arithmetic of rational numbers; this means that it cannot prove the laws of this arithmetic as truths. However, such experiences could very well incite the mathematician to create a calculus which is applicable to these underlaid objects. We will frequently return to this point of view.

2. Why do we use number couples rather than, for example, number triples? The answer is that our calculus is copied after the arithmetic of quotients, and a quotient is completely determined by specifying two numbers. (And the same is true for the integers, whose calculus is copied after the arithmetic of differences.) The fact that inverse operations

5. THE RATIONAL NUMBERS

always apply to *two* numbers is a more basic reason why we just need number couples.

A number couple is nothing in itself; it is an empty frame which can be filled with the most varied content. Thus, the same number couple will represent, according to the rules that are set up, an integer, or a rational number, or a complex number. If we wish to speak of the meaning of a number couple, the best we can say is that the meaning depends on the kind of application.

3. In the construction described above, the integers were first constructed and then the rational numbers. Is there an innate necessity for this sequential order? Couldn't we first introduce the rational numbers without signs and then the distinction between positive and negative numbers? Certainly! We would not thereby obtain another system of rational numbers; rather, the system so constructed would prove to be isomorphic to the one considered above, since every relation of one system could be mapped on a similarly constructed relation in the other, and conversely.

6. Foundation of the Arithmetic of Natural Numbers

Our entire development is based on the system of natural numbers. Hitherto we have assumed that they are known or given; we must now seek a deeper insight into these numbers. For this purpose we will trace the source of the laws of calculation themselves. We will thereby be confronted with issues which even today have not been settled. We will first delineate the present status of these investigations and then develop some ideas which may help to clear up some point or other. Calculations with natural numbers can be based on a few laws which we first wish to list synoptically.

I. *Laws for addition*

1. $a + b$ is always a number, that is, addition can be performed without restriction.
2. $a + b$ is uniquely determined, that is, there is only one number which is the sum of a and b.
3. $a + b = b + a$ (commutative law).
4. $a + (b + c) = (a + b) + c$ (associative law).
5. If $a > b$, then $a + c > b + c$ (law of monotony).

II. *Laws for multiplication*

6. $a \cdot b$ is always a number.
7. $a \cdot b$ is uniquely determined.
8. $a \cdot b = b \cdot a$ (commutative law).
9. $a \cdot (b \cdot c) = (a \cdot b) \cdot c$ (associative law).
10. If $a > b$, then $a \cdot c > b \cdot c$ (law of monotony).

6. Foundation of the Arithmetic of Natural Numbers

IV. *Laws uniting addition and multiplication*

11. $a \cdot (b + c) = a \cdot b + a \cdot c$ (1st distributive law).
12. $(a + b) \cdot c = a \cdot c + b \cdot c$ (2nd distributive law).

(Later we will discuss the laws of calculation for the inverse operations.)

Let us first examine how these laws can serve as a basis for elementary computations. We will illustrate this by the multiplication of 7 and 24. For this purpose we separate 24 into 20 and 4 and compute
$$7 \cdot 24 = 7 \cdot (20 + 4) = 7 \cdot 20 + 7 \cdot 4,$$
thereby using the distributive law. In computing
$$7 \cdot 20 = 7 \cdot (2 \cdot 10) = (7 \cdot 2) \cdot 10 = 14 \cdot 10 = 140$$
the associative law of multiplication is used; on separating 28 into $20 + 8$, and computing
$$140 + 28 = 140 + (20 + 8) = (140 + 20) + 8 = 168$$
the associative law of addition is used.

Thus, in the individual steps of computing with natural numbers we recognize exactly our general laws. What role is played by the monotonic laws? They are not used in ordinary computations but are important in abridged multiplication, where it is a question of enclosing the result between two fixed limits. To summarize: practical computations with numbers consist in the repeated application of these twelve basic laws, wherein the results for the digits can be taken from a memorized set of relations (the addition and multiplication tables).

Naturally the practice of computing had been in effect long before these laws were known explicitly; they had first to be picked out of the computational processes and be recognized as their logical foundation. This happened in the first third of the last century, mainly through the efforts of English and French mathematicians (such as Hamilton and Servois). These ideas first penetrated into Germany in 1867 through Hankel and then later through Stolz.

The following question immediately puts us in the midst

of the modern controversy. How can these twelve laws be justified? Can they be proved or must they be taken as undemonstrable basic truths? Opinions differ regarding these matters. We can distinguish the following four points of view.

I. The basic laws are truths which are self-evident to any thinking mind. For instance, from the intuitive meaning of addition as a combination of two sets we read immediately that there can be only *one* number which can serve as their sum, and that this result is independent of the order in which the two sets are combined. From this point of view it is intuition which is the source of all mathematical knowledge, where it is to be understood to refer more to the inner perception of the number series than to the physical experience of empirical things. Among the philosophers, Kant is a representative of this view; while among the mathematicians, Hamilton can be mentioned.

II. We can try to reduce the number of basic laws by deriving them from fewer but deeper-lying propositions. This point of view is a modification of the first one. It amounts to saying that only the primary laws which the investigation uncovers shall be taken from intuition, while all others shall be deduced from these by the rules of logic. H. Grassmann prepared the way in this direction. He showed in his *Lehrbuch der Arithmetik* (*Manual of Arithmetic*), 1861, that the commutative law can be derived from the associative law by means of the principle of complete induction. The conclusion of this development is marked by Peano. According to Peano the whole structure of arithmetic can be built from five basic propositions. If the truth of these propositions is admitted, then all remaining ones follow as purely logical conclusions, without any further reference to the intuitive significance of the arithmetical operations. These investigations are extraordinarily laborious. The principal difficulty is to keep aloof from the countless associated ideas which arise from the use of ordinary language and which threaten to enter unnoticed into the dis-

6. FOUNDATION OF THE ARITHMETIC OF NATURAL NUMBERS

cussion; otherwise the conclusions, which should be drawn only from the explicitly given premises, would be robbed of their strength. To facilitate their control, Peano invented a symbolic language by which he formalized the logical conclusions so as to be sure that an assumption should not enter unnoticed in a chain of inferences. Thus the beginnings of modern logic arose from the requirements of the mathematical technique of proof. The basic propositions of Peano are:

1. Zero is a number.
2. The successor of any number is another number.
3. There are no two numbers with the same successor.
4. Zero is not the successor of a number.
5. Every property of zero, which belongs to the successor of every number with this property, belongs to all numbers.

Later we will return to the precise meaning of these propositions. Here we only remark that these investigations did not find the ultimate basis of mathematical knowledge but only pushed the fundamental principles further back. Therefore efforts were exerted toward deducing the basic propositions of Peano from deeper-lying truths. Any foundation of these truths with the help of arithmetic is out of the question; for in the Peano axioms we actually have already reached the last starting point of arithmetical deductions. But perhaps such a possibility may be disclosed if we examine the frontier of arithmetic. This leads us to the third point of view whereby we try

III. to base arithmetic on logic. For this purpose we use very general formations of concepts from the theory of sets, or from the calculus of classes. The originator of this school of thought was Frege. While Peano brought the arithmetization of mathematics to its highest point, its logicalization was started by Frege. The assertion that mathematics is only a branch of logic contains two distinct theses within itself, which are not always clearly kept apart;

6. Foundation of the Arithmetic of Natural Numbers

a) The basic concepts of arithmetic can be reduced by definition to purely logical concepts.

b) The basic propositions of arithmetic can be deduced by proofs from purely logical propositions.

If Frege's assertion is correct, it would mean that the totality of pure mathematics—inasmuch as it is based on the natural numbers—has the same character as logic. At this point, it is perhaps appropriate to say something about the nature of the propositions of logic. Philosophers frequently discuss the question: what does the validity of logic depend on? Does the intrinsic truth of its propositions force itself on any thinking mind, or is it due to the empirical nature of our consciousness, etc.? Today we answer that the propositions of logic are tautologies.

What is a tautology? It can be fairly well illustrated by the following example. Let us assume that someone does not exactly know what day of the week we have just written. Perhaps he would answer our question: today is Monday or Tuesday. This would not be a very definite answer. It would be still more indefinite if he wavered between three days. The answers can be sequentially arranged according to their indefiniteness—today is Monday, today is Monday or Tuesday, today is Monday or Tuesday or Wednesday, etc.—whereby each statement has less sentential content than the previous one. The one which says "Today is Monday or Tuesday ... or Sunday," is the most indefinite statement of this kind; it allows reality the greatest possible latitude. However, this proposition has a very noteworthy property which basically distinguishes it from all preceding ones, namely, it can not be false. Whatever day of the week we may have written, it must be either Monday or Tuesday, etc. Such a proposition is true because of its mere form and tells us nothing about reality. A formation of this kind is called a tautology. Another example is the proposition: "Today either I go out or I do not go out"; for, whatever I do, the truth of the statement is irrefutable. (We readily notice that this ex-

6. Foundation of the Arithmetic of Natural Numbers

ample has the form of the law of the excluded third, which is also a tautology; and the same is valid of all remaining propositions of logic.) Hence a tautology is a proposition which is true because of its mere form but whose truth is bought at the price of a complete lack of content.

If Frege's opinion proves correct, it would mean that mathematics represents an immense system of tautologies, and therefore says absolutely nothing. Strangely enough, Frege failed to realize that *this* was the consequence of his theory; for he still lacked entirely insight into the nature of the logical. In his opinion logic should be a descriptive science, such as mechanics; to the question, what does it describe, he would answer: relations between ideal objects as "and," "or," "if," etc. According to this conception there is a domain of logical structures which is not created by the human mind and reigns as timelessly as the ideas of Plato; between these there exist many relations, which logic investigates. In this way Frege constructed his system of logic believing that in every case he was advancing to new and always deeper truths. The discovery that the propositions of logic are no more than tautologies has corrected this view. At any rate, it has given an interpretation to the logicalization of mathematics which its author would hardly have approved.

These trains of thoughts belong, however, to a later date.[1] The disaster broke in from another quarter. Just as Frege had concluded the second volume of his *Grundgesetze* (*Basic Laws*), after ten years of work, he received a letter which disclosed that one of his syllogisms led to an antinomy. The writer of this letter was Bertrand Russell. Frege had to recognize, at the very moment when he believed his work to be completed, that he had built on sand, or at least that the foundations must be renewed. In an epilogue he communicated this fact to the reader, and then actually suspended further work.

[1] The concept of tautology was introduced by Wittgenstein in the year 1921.

6. Foundation of the Arithmetic of Natural Numbers

With the discovery of antinomies the development of logic entered a new phase. Hereafter efforts were aimed at clearing up this remarkable phenomenon. It was recognized that antinomies arose from handling the concept of set too loosely and incautiously. We wish to illustrate this point by an example. Either a set contains itself as an element or it does not contain itself. The set of all men is not a man, the set of all points is not a point; however, the set of all abstract concepts is itself an abstract concept. We will call a set normal if it does not contain itself as an element. (The first two examples belong to this category.) Let us think of all normal classes as combined into a new set N. We now ask whether or not N is normal, that is, contains itself as an element. Let us first assume that N contains itself as an element; then the set N occurs among its proper elements. Consequently N contains a non-normal set, N itself—whereas by definition N contains only normal sets. This contradiction shows that our assumption was false. Consequently, we will say that only the contrary statement can be correct. However, the startling thing is that the contrary statement also leads to a contradiction. If N does not contain itself as an element, then N is a normal set; however, since N contains all normal sets, it must also contain the normal set N, that is, it contains itself—again a contradiction. The crux of the whole matter is that the antinomy, if we trace its genesis, goes back directly to the concept of set, so that the source of the contradiction must lie in the formation of this concept. Hence Russell attempted to confine the concept of set, and he did this by stating certain restrictions. Accordingly, we can no longer indiscriminately combine things, sets of things, sets of sets, etc. Instead we must be careful that the elements of a set have a certain homogeneity. This is the basic idea of the so-called theory of types—a theory which we cannot go into here. This theory has actually barred the path of the antinomies known up to now. However, it does not give an absolute guarantee. For who will vouch that some day new antinomies will not emerge?

6. Foundation of the Arithmetic of Natural Numbers

To use the words of Poincaré: Haven't we perhaps unwittingly also enclosed the wolf in the pen we have built for the sheep of the theory of sets?

IV. As a consequence the reduction of mathematics to logic lost much of its value in the eyes of mathematicians. Rather it seemed necessary to place the consistency of the new logic on a firm footing through mathematical investigations. Since, however, in mathematics itself logical ways of inference are again used, the conjecture becomes more and more plausible that the end must be sought in another direction, namely, in a common construction of mathematics and logic. Let us consider what this means. We no longer ask how the consistency of an individual theory is to be proved, for example, that of non-Euclidean geometry; instead the much bigger problem presents itself, namely, to prove the consistency of arithmetic and logic simultaneously.

The method used in these investigations is the axiomatic one. It was first applied in geometry and then gradually forced its way into other domains. Its nature can be very well characterized by the following comparison.

According to the older view the axioms describe facts of immediate intuition. They deal with the "ideal" points, lines, planes and those relations, which are characterized by the words "incident," "congruent," "between," "parallel." Accordingly Euclid begins with the definition of the basic concepts; as he puts it: σημεῖον, οὗ μέρος οὐδέν, point is that which has no parts. These definitions formed a stumbling block from the very beginning, for their meaning is extraordinarily obscure. For example, a pain has no parts; now is a pain a point? However, we must say, above all, that even if such a definition were understood strictly speaking it would have no value for Euclid's system. Not a single proof depends on its explanation; it is never used; it stands entirely outside of the remaining system of propositions. Now the objective of the definition is to create concepts which are starting points of sharp deductions;

this end is not realized.[2] Furthermore, another fact stands out. In modern geometry mathematicians came to realize that the geometrical propositions of one subject could be "transferred" to an entirely different one. For example, all propositions which are valid for the straight lines of our three-dimensional space can be interpreted as dealing with the points of a four-dimensional space. The two systems of concepts are completely isomorphic (similarly constructed); if a proposition had been proved in the geometry of one manifold, it could be automatically transferred and, without further proof, immediately stated as a proposition in the geometry of the other manifold.[3] *Hence the perceptual appearance of the basic geometrical elements has nothing whatever to do with the validity of the propositions.* The reader has met another example of a transfer in the case of the Euclidean model of the non-Euclidean geometry of Bolyai. This situation has led the mathematician to loosen the logical framework of a theory from its intuitive or empirical content. We no longer explicitly define a point, a line, a plane, or explain the meaning of the basic relations. Instead we first lay down axioms which characterize each elementary concept in its totality, and then we say that points, lines, and planes are things satisfying the postulated axioms. According to Pieri, such a theory must be understood as a hypothetical-deductive system. This means that if the assump-

---

[2] Only lately has a geometric system of axioms been found in which the concept of part is rigorously defined. In this system points are defined as those things which have no parts except for themselves and the "empty thing," and from this definition a number of inferences are deduced. Surely this example shows rather clearly the deficiencies of Euclid's definition. The system of axioms just mentioned was given by Menger in *New Foundations of Projective and Affine Geometry*. Annals of Mathematics, 1936.

[3] Another beautiful example of such a transfer is found in Schlick, *Allgemeine Erkenntnislehre* (*General Theory of Knowledge*), Section 7.

6. FOUNDATION OF THE ARITHMETIC OF NATURAL NUMBERS

tions prove to be correct in some domain, i.e., if we can produce objects which behave in accordance to the relations demanded by the axioms, then all theorems deducible from them are also true. Every such assignment of objects leads to a definite meaning or realization of the system of axioms.

Even the limitation to spatial relations is not necessary. Hilbert gave a drastic example of this. "Drosophila is a small but very interesting fly. It has been the object of the most extensive, careful and successful breeding experiments. This fly is usually grey, red-eyed, spotless and has long round wings. However, there also occur flies with special irregular characteristics: instead of grey they are yellow, instead of having red eyes they have white eyes, etc. Usually these five special characteristics are tied together, that is, if a fly is yellow, it also has white eyes, is spotted and has cloven club wings. And if it has club wings it is also yellow, has white eyes, etc. Besides these usual correlations, slight deviations appear among the descendants, resulting from suitable cross-breedings. However, percentagewise the number of offsprings with like characteristics is always the same. These numbers, which are found experimentally, satisfy the linear Euclidean axioms of congruence and the axioms concerning the geometrical concept of betweenness.' Thus the laws of heredity appear as an application of the linear congruence axioms, that is, of the elementary geometrical propositions concerning the laying off of line segments; simple and precise—and at the same time wonderful beyond any stretch of the boldest imagination."[4]

We can see that in geometry the axiomatic method results in a complete separation of the logical-formal from the spatial-intuitive. We can apply this procedure to other domains, for example, to arithmetic. The structure of such a formal construction would be as follows: we must first classify the concepts

[4] *Naturerkennen und Logik*, Naturwissenschaften (*Knowledge of Nature and Logic*, Science), 1930.

6. FOUNDATION OF THE ARITHMETIC OF NATURAL NUMBERS

and propositions of arithmetic according to whether or not they are needed for the logical construction of the theory. The propositions which are suitable for this purpose are called axioms. We need not know what the axioms mean; instead we restrict ourselves to the problem of deducing other propositions from these axioms. In the propositions and syllogisms we are confronted with concepts which are rendered by the words "and," "or," "if," "not," "all," "there is," etc.; we usually assume that we know the precise significance of these words and how to handle them by the laws of logic. Now the very objective of the formal construction should be to prove the consistency of arithmetic and logic. Hence we ought not to use logical conclusions in the derivation of arithmetical formulae, if we wish to avoid a vicious circle. Hilbert took the following stand on this point: arithmetic consists of propositions and discussions; the former are expressed by a symbolic language, the latter, by words. (We only have to think of a mathematical textbook in which formulae and text alternate with one another.) We now extend this conception by including the discussions in the structure of formal arithmetic; accordingly, these must also be expressed as formulae. The previous development of logical calculus (Peano, Frege, Russell) had already assured us that the "intuitive conclusions" could be imitated symbolically. And so we take the last step by also abandoning the intuitive significance of the logical concepts. Thereby the symbols of the logical calculus become marks without content just as previously the words "point," "line," "plane," "congruent," etc., were. We only assume that certain combinations of these symbols, the axioms of logic, exist, which we think of as added to the axioms of arithmetic.

We could compare mathematics so formalized to a game of chess in which the symbols correspond to the chessmen; the formulae, to definite positions of the men on the board; the axioms, to the initial positions of the chessmen; the directions for drawing conclusions, to the rules of movement; a proof, t

6. Foundation of the Arithmetic of Natural Numbers

a series of moves which leads from the initial position to a definite configuration of the men. Naturally this does not mean that mathematics is "only a game" (that "one thereby thinks of nothing"); instead it merely means that for the purpose of a very definite investigation—the proof of consistency—the intuitive significance can be disregarded. To every concept of the meaningful mathematics there corresponds a formula in our game of proofs, and to every significant discussion, a sequence of such formulae having definite and precisely stated characteristics. And therefore a picture of mathematics can be thought of as formed in the domain of symbolism. We can now study this system of formulae itself and ask whether it is consistent. Since our formulae do not express ideas, our question does not have as yet a clear meaning. We must first define the property "consistency," for example, through the postulate: This property is attributed to a system of formulae only if the proofs cannot lead to the formula $1 \neq 1$.

Besides this formalized mathematics there arises one which has content, namely, metamathematics. It investigates the structure of the formalized mathematics; its primary aim is to attain an insight into the concept of consistency. From the point of view just described, we cannot ask, at all, regarding the "truth" of the axioms. However, metamathematics brings to light real intuitive knowledge. "The axioms and demonstrable propositions," explains Hilbert, "are the images of ideas which constitute the usual procedures of the mathematics so far developed. However, they are not themselves truths in the absolute sense. We have to regard as absolute truths rather the insights, furnished by my theory of proof, concerning the demonstrability and consistency of those systems of formulae."[5]

The arguments of this metamathematics shall always be of a finite and intuitive nature. The subject matter of this theory

---

[5] *Die logischen Grundlagen der Mathematik* (*The Logical Foundations of Mathematics*) Math. Ann. 88.

6. FOUNDATION OF THE ARITHMETIC OF NATURAL NUMBERS

are the symbols themselves, which "can be surveyed perfectly in all parts," and whose identification, distinction, sequence "is intuitively evident to us simultaneously with the objects, as something which can no longer be reduced to anything else."[6] The kind of discussions that present themselves in this theory may be illustrated by two examples:

1. In a demonstrable formula let a definite symbol occur more than twice. Then, if we check the proof, we must encounter a formula which possesses this property for the first time.

2. If, in a finite sequence of formulae, the first has a certain property which can always be transferred from a formula to its successor, then *all* formulae have this property. (Finite induction.)

It is evident that certain facts of intuition are formulated here—simple properties of ordering, which can be described satisfactorily by a kind of rudimentary number concept—which are so unquestionable and self-evident that they precede every application of logical conclusions. There seems to be a kind of primitive system, which supports the entire structure of logic and arithmetic; if these facts are not admitted, we may as well cease to think and try to understand. Similar aims had already been pursued earlier by König in his text *Neue Grundlagen der Logik, Arithmetik und Mengenlehre* (*New Foundations of Logic, Arithmetic and Theory of Sets*). We will report later on the results of these attempts.

---

[6] *Neubegründung der Mathematik*, Abhandlungen aus dem Mathematischen Seminar der Hamburger Universität (*Refoundation of Mathematics*, Papers from the Mathematical Seminary of Hamburg University) 1922.

7. Rigorous Construction of Elementary Arithmetic

After having briefly surveyed the various standpoints, we now undertake a more exact discussion of the individual views.

Our first question is: What does a rigorous construction of elementary arithmetic look like? Naturally in such a system tacit assumptions or undefined concepts cannot be used—except for the basic concepts. We wish to exhibit for the reader the basic outlines of such a construction. For this purpose we will follow the representation of Skolem.[1] The aim of this work is to deduce the basic formulae of arithmetic (our basic laws 1 to 12) from mere definitions using only the principle of complete induction.

Our twelve basic laws describe certain properties of addition and multiplication. As yet we have not exactly defined what is to be understood by addition and multiplication, but have assumed that these operations are known from school instruction. In a rigorous procedure this is not allowable, and therefore our first task is to define the four basic operations clearly. Every definition assumes other concepts to which it reduces those to be explained. Hence we must indicate the concepts which will be the basis of our investigation. We think of the following as undefined basic concepts:

1. The concept of "natural number";

---

[1] *Foundation of Elementary Arithmetic*, Skrifter Videnskapsselkapet i Kristiania, 1923.

7. Rigorous Construction of Elementary Arithmetic

2. the concept of "successor" or "the number n + 1 following the number n";

3. the concept of equality, namely, "a = b" shall mean that a can be replaced by b and conversely.

Furthermore, the principle of complete induction will be used in the proofs. (We will discuss later the nature of this principle.)

What shall we now understand by the sum of two numbers a + b? An interpretation familiar to the reader is that the sum is the number which is obtained by adding 1 to a as often as b indicates. In this way addition is reduced to the repeated application of the operation + 1. This loosely stated concept will now be put into a precise form.

DEF. 1. $\qquad a + (b + 1) = (a + b) + 1.$

What does this definition mean? Perhaps it can be best characterized by saying that it is a direction for the formation of definitions, that is, it is the general rule by which the individual definitions

$$a + 2 = (a + 1) + 1$$
$$a + 3 = (a + 2) + 1$$
$$a + 4 = (a + 3) + 1$$

etc.

are formed. Hence, if I only know the meaning of a + 1, (that is, what is to be understood by the successor of a), then the first definition tells me what a + 2 means; if I know what a + 2 indicates, then by the second definition I know the meaning of a + 3, etc.; and the generic definition is the prototype of all these individual definitions. Such a definition is called a *recursive definition*; it is the general term of a series of rules.

Perhaps the reader is of the opinion that we are following a vicious circle; for, if I use the plus sign in my explanation am I not already assuming the concept of sum? Not at all. Up to now we know the sign "+" only in the combination

7. RIGOROUS CONSTRUCTION OF ELEMENTARY ARITHMETIC

a + 1. We now wish to interpret it in general. For this purpose we give a direction, according to which the expressions a + 2, a + 3, ... a + b, a + (b + 1) are to be formed, by saying:

$$a + (b + 1) = (a + b) + 1.$$

This means that the sum of a and b + 1 can always be replaced by the number following a + b. If addition is already defined for arbitrary values of a and for a definite number b, then this definition explains the sum of an arbitrary a and b + 1; consequently, addition is defined in general.

In the following we will prove a number of propositions which are all known to the reader and which perhaps seem entirely self-evident to him. Many a person will think that it is downright superfluous to prove something in minute detail which no one doubts anyway. If the reader should share this view, we ask him to bear in mind that the objective of the following proofs is not to evoke in us a feeling of conviction, but rather to attain an insight in the dependence of the propositions. The requirement, that everything should be proved which can be proved, does not descend from a skeptical disposition; instead it is only the expression of the desire to see clearly the structure of the complex propositions and to recognize what combinations the individual truths of arithmetic have among one another. The reader should appraise the following proofs from this point of view.

THEOREM 1. $a + (b + c) = (a + b) + c$ Associative law.

PROOF: By Def. 1 the theorem is valid for c = 1. Let us assume that it is valid for a definite c and every value of a and b. We will now show that it is also valid for c + 1 and every value of a and b, i.e., the formula

$$a + (b + [c + 1]) = (a + b) + (c + 1)$$

is true. This will be proved by showing that the expression on the left-hand side of the equality sign can be transformed until it takes on the form of the expression on the right-hand side;

7. RIGOROUS CONSTRUCTION OF ELEMENTARY ARITHMETIC

in doing this we will clearly set off the individual steps of the transformation.

By Def. 1 we have $b + (c + 1) = (b + c) + 1$; it follows that

(α) $\quad a + (b + [c + 1]) = a + ([b + c] + 1)$;

by Def. 1 the right side is also equal to

(β) $\quad a + ([b + c] + 1) = (a + [b + c]) + 1$;

but by assumption $a + (b + c) = (a + b) + c$, therefore

(γ) $\quad (a + [b + c]) + 1 = ([a + b] + c) + 1$.

By Def. 1 we finally have

(δ) $\quad ([a + b] + c) + 1 = (a + b) + (c + 1)$.

From these four transformations it now follows that the first and last terms of the chain must be equal; we therefore obtain

$$a + (b + [c + 1]) = (a + b) + (c + 1)$$

which is what we wanted to prove. The structure of the proof can be described as follows: first the theorem is shown to be valid for $c = 1$; then it is proved that, if the theorem is valid for c, it must also be valid for $c + 1$; and from this it is concluded that the theorem is valid in general. Inferences of this type are called inferences by *complete induction*. By virtue of this inference the associative law is an immediate consequence of the definition of sum. We note in passing that the principle of complete induction is not a premise from which inferences are drawn, but a method for arriving at inferences.

From the above theorem it follows that parentheses can be deleted in an expression such as $a + (b + c)$. Furthermore the associative law is the basis which permits the extension of the concept of the sum of two numbers to that of three or arbitrarily many numbers.[2]

The proof of the commutative law of addition will be based on the following

LEMMA: $\qquad a + 1 = 1 + a$

[2] The reader who is not primarily interested in mathematics may omit the rest of this chapter.

PROOF: The lemma is valid for $a = 1$ (for $1 + 1 = 1 + 1$). Let us assume that the lemma is valid for a. We will now show that it must also be valid for $a + 1$. By our assumption we have $a + 1 = 1 + a$; therefore
(α) $\qquad (a + 1) + 1 = (1 + a) + 1;$
by Def. 1 we also have
(β) $\qquad (1 + a) + 1 = 1 + (a + 1),$
from (α) and (β) it follows that
$$(a + 1) + 1 = 1 + (a + 1),$$
which states that the lemma is valid for $a + 1$.

THEOREM 2. $\qquad \underline{a + b = b + a} \qquad$ Commutative law.

PROOF: As a consequence of the lemma theorem is valid for $b = 1$. Let us assume that it is valid for b. We will now show that it is also valid for $b + 1$, which means that
$$a + (b + 1) = (b + 1) + a$$
is valid.

We will again separate the proof in a series of individual steps:

(α) $\quad a + (b + 1) = (a + b) + 1 \qquad$ (Def. 1)
(β) $\quad (a + b) + 1 = (b + a) + 1 \qquad$ (by assumption)
(γ) $\quad (b + a) + 1 = b + (a + 1) \qquad$ (Def. 1)
(δ) $\quad b + (a + 1) = b + (1 + a) \qquad$ (Lemma)
(ε) $\quad b + (1 + a) = (b + 1) + a \qquad$ (Theorem 1).

From this chain of five equations it follows that
$$a + (b + 1) = (b + 1) + a.$$

It should be mentioned that in this construction the associative law was the first to be introduced because it is needed (transformation ε) in the proof of the commutative law.

We will now see how to obtain further concepts of arithmetic. How shall we define, let us say, the concept "smaller"? If the reader were to propose a definition, he would perhaps say that a is smaller than b if something must be added to a in order to obtain b; in other words, if there is a number x such that
$$a + x = b.$$

7. Rigorous Construction of Elementary Arithmetic

This formulation is not false, but it is burdened with a characteristic drawback. Namely, it contains an appeal to an infinite—and therefore unaccomplishable—task, because the criterion of being smaller consists in finding, by trial and error, among all possible numbers a number x such that $a + x = b$. A definition that achieves the same end while avoiding the appeal to the nonending number series would be preferable. This can actually be done by defining the concept "smaller" by a recursive definition. For this purpose we assert

DEF. 2.

$a < 1$ is always false;
$a < b + 1$, if either $a < b$ or $a = b$.

First, this definition explains that $a < 1$ is never valid. Second, it establishes that the relation "smaller" is valid between an arbitrary a and a particular $b + 1$ whenever this relation is already defined for b. As can be seen, this definition no longer involves the concept of existence.

We now explain the relation "greater" as the converse of the relation "smaller":

DEF. 3.

$a > b$ means $b < a$.

From these definitions various theorems can be derived which we will cite here without proof:

3. The relation "smaller" is transitive, that is, this means if $a < b$ and $b < c$, then $a < c$.

4. It is inflexive, that is, a number is never smaller than itself.

5. The relations "smaller," "equal," "greater" form a complete disjunction, that is, if a and b are two arbitrary numbers, either $a < b$ or $a = b$ or $a > b$ is valid.

6. If $a < b$, then $a + c < b + c$ is also valid.

7. From $a + c = b + c$ follows $a = b$; for the particular case $c = 1$ the theorem shows that any number has only *one* predecessor.

7. RIGOROUS CONSTRUCTION OF ELEMENTARY ARITHMETIC

How can multiplication be defined? The procedure described above suggests that here also we should try a recursive definition. Hence we will first explain how to multiply by 1 and then reduce the multiplication by $b + 1$ to the multiplication by b.

DEF. 4.
$$a \cdot 1 = a$$
$$a \cdot (b + 1) = a \cdot b + a.$$

Thereby we have again given a direction for the formation of definitions. Thus from the first part of the definition I am told what $a \cdot 1$ means, namely, a. The second part tells me that $a \cdot 2$ means the same as $a + a$, $a \cdot 3$ just as much as $a \cdot 2 + a$, etc. In other words, we have given the general rule according to which the individual definitions are formed.

We have now come to a point where we can prove some more of our basic laws.

THEOREM 8. $\underline{a \cdot (b + c) = ab + ac}$ (First distributive law).

PROOF: The theorem is valid for $c = 1$ (by Def. 4). Let us assume that the theorem is valid for a certain c and for any value of a and b. We wish to show that it will also be valid for $c + 1$ which means that
$$a \cdot (b + [c + 1]) = ab + a \cdot (c + 1).$$

Now

(α)	$a (b + [c + 1]) = a ([b + c] + 1)$	(Def. 1)
(β)	$a ([b + c] + 1) = a (b + c) + a$	(Def. 4)
(γ)	$a (b + c) + a = (ab + ac) + a$	(by assumption)
(δ)	$(ab + ac) + a = ab + (ac + a)$	(Theorem 1)
(ε)	$ab + (ac + a) = ab + (a[c + 1])$	(Def. 4)

From (α), (β), (γ), (δ) and (ε) follows
$$a (b + [c + 1]) = ab + a (c + 1).$$

THEOREM 9. $\underline{a (b \cdot c) = (ab) \cdot c}$ (Associative law).

PROOF: as above.

THEOREM 10. $\underline{(a + b) c = ac + bc}$ (Second distributive law).

PROOF: as above.

7. Rigorous Construction of Elementary Arithmetic

LEMMA: $\qquad 1 \cdot a = a \cdot 1$

PROOF: as above.

THEOREM 11: $\qquad a \cdot b = b \cdot a \qquad$ (Commutative law).

PROOF: as above.

Further theorems which can be proved are:

THEOREM 12. If $a < b$, then $ac < bc$ and conversely.

THEOREM 13. $\qquad a < ab$.

Before defining subtraction and division we must say something about the concept of divisibility. How shall we express that a number a is divisible by a number b? The first thought which comes to us is to say that a is divisible by b if there is a number x such that $a = bx$. This again involves an appeal to the nonending number series. However it is easy to free oneself from the infinite. If there is to be a number x with the required property, it must be one of the numbers 1, 2, ... a. (Theorem 13.) This permits us to define the relation of divisibility as follows:

D (a, b) means that $a = b$ or $a = 2b$ or $a = 3b$ or ... $a = ab$.

Consequently, the statement "a is divisible by b" can be split into a finite disjunction so that the validity or non-validity of this statement can be substantiated in finitely many steps. The finite disjunction itself can be defined by recursion. For this purpose we will take a seemingly roundabout way. We first define a three-place relation Δ (a, b, 1), which shall mean that a is equal to b multiplied by a number which lies between 1 and c, and then explain the divisibility D in terms of this concept Δ. The exact definitions are the following:

DEF. 5.

$\qquad \Delta$ (a, b, 1) means $a = b$,

Δ (a, b, c + 1) means Δ (a, b, c) or $a = b(c+1)$.

DEF. 6.

$\qquad D (a, b) = \Delta (a, b, a).$

7. Rigorous Construction of Elementary Arithmetic

This definition allows us to prove various theorems. For example, if a is divisible by b and b is divisible by c, then a is also divisible by c; also, if both a and b are divisible by c, then $a + b$ is also divisible by c. Regarding the details of the proofs, we refer to the treatise of Skolem. We will now consider subtraction and division.

Subtraction can be defined as follows:

DEF. 7. $\quad c - b = a$ means $c = a + b$.

The expression $c - b$ is called a difference. It can be proved that this expression represents a natural number only if and that in this case it represents only *one* number. However, the proof is somewhat lengthy so that it is omitted.

Division can be defined in a manner analogous to that of subtraction.

DEF. 8. $\quad \dfrac{c}{b} = a$ means $c = a \cdot b$.

The expression $\dfrac{c}{b}$ is called a quotient. As before, we can prove that $\dfrac{c}{b}$ represents a natural number only if $D(c, b)$ is valid; and if $D(c, b)$ is valid, then $\dfrac{c}{b}$ has only one value. On this account $D(c, b)$ is completely equivalent to the statement of the existence of a value of $\dfrac{c}{b}$ (between 1 and c).

We conclude by citing a number of theorems which are immediate consequences of the preceding.

$$(a - b) + b = a \qquad \frac{a}{b} \cdot b = a$$

$$(a - b) + c = (a + c) - b \qquad \frac{a}{b} \cdot c = \frac{ac}{b}$$

$$(a - b) - c = a - (b + c) \qquad \frac{\left(\frac{a}{b}\right)}{c} = \frac{a}{bc}$$

$$a - (b - c) = (a - b) + c \qquad \frac{a}{\left(\frac{b}{c}\right)} = \frac{a}{b} \cdot c$$

$$(a - b) \cdot c = ac - bc$$

$$\frac{a + b}{c} = \frac{a}{c} + \frac{b}{c} \qquad \frac{a - b}{c} = \frac{a}{c} - \frac{b}{c}$$

8. The Principle of Complete Induction

In the construction of arithmetic, as delineated in the previous chapter, the principle of complete induction occupied a commanding position. By means of this procedure all basic laws were deduced from mere definitions. However, what can we say about complete induction itself? Even today the genesis of this principle has not been entirely cleared up. We will illustrate this by comparing the views of a most eminent mathematician with those of a most eminent logician living at the turn of the century, namely, Poincaré and Frege.

According to Poincaré the achievement of the inductive method is that it contains, condensed, so to speak, in a single formula, an infinity of syllogisms. In order to see this better, let us formulate the syllogisms one after another. We first have that theorem A is true of the number 1. Now if it is true of the number 1, it is also true of the number 2; therefore it is true of 2. Now if it is true of the number 2, it is also true of 3; consequently it is true of 3, etc. In symbols

$$\frac{\begin{array}{c} A\,(1) \\ A\,(1) \rightarrow A\,(2) \end{array}}{A\,(2)}$$
$$\frac{A\,(2) \rightarrow A\,(3)}{A\,(3)}$$
. . . .

The individual syllogisms follow one another like the cascade of a waterfall. The result of any one conclusion is used as the minor of its immediate successor. Poincaré continues:

8. The Principle of Complete Induction

"If, instead of showing that our theorem is true of all numbers, we only wish to show it true of the number 6, it will suffice for us to establish the first five syllogisms of our cascade; nine would be necessary if we wanted to prove the theorem for the number 10; more would be needed for a larger number; but however great this number might be, we should always end by reaching it, and the analytical verification would be possible.

"And yet, however far we might go in this manner we would never rise to the general theorem, applicable to all numbers, which alone can be the object of science. To reach this, an infinity of syllogisms would be necessary; it would be necessary to overleap an abyss, which the patience of the analyst, restricted to the resources of formal logic alone, never could fill up.

"Therefore, we cannot escape the conclusion that the rule of reasoning by recurrence is irreducible to the principle of contradiction." The latter principle "would always permit us to develop as many logical conclusions as we wished. It is only when it is a question of including an infinity of them in a single formula, it is only before infinity that this principle fails, and here, too, experience becomes powerless. This rule, which is just as inaccessible to analytic demonstration as it is to experience, is the veritable type of the synthetic a priori judgment. On the other hand we cannot think of seeing in it a mere convention as in some of the postulates of geometry."[1]

Frege, on the contrary, held the view that by his investigations "it has been made plausible that the arithmetical laws are analytic judgments and consequently a priori. According to him, arithmetic is only an extended logic, every arithmetical proposition is a law of logic, although a derived one.

"Kant has underrated the value of analytical judgments... The more fruitful definitions of concepts draw boundary lines

[1] *The Foundations of Science*, 1, Chap. I.

8. THE PRINCIPLE OF COMPLETE INDUCTION

which were not yet given at that time: what I am allowed to conclude from them, is not to be ascertained beforehand; it is not simply a question of drawing out of a box what has been put into it. These conclusions extend our knowledge and therefore, according to Kant, should be considered as synthetic; yet they can be proved by purely logical methods and are therefore analytic. To be sure they are contained in the definitions, but as a plant in a seed, not as the beams in a house. We often need several definitions for the proof of a proposition. Even though in such a case it is not contained in one definition, it still follows by purely logical methods from all of them taken together."[2]

The attitude of these investigators toward Kant is remarkably ambivalent. Poincaré is of the opinion that Kant was mistaken in his estimation of the propositions of geometry. According to Poincaré, these are analytic, for the axioms are basically only conventions. On the other hand, Poincaré agrees with Kant that in the domain of arithmetic synthetic a priori judgments actually occur. Frege holds the view that the proposition of arithmetic are analytic. However, he agrees with Kant regarding the domain of geometry: when Kant in this domain "calls truths synthetic and a priori, he has disclosed their true substance."

Poincaré agrees with well-known mathematicians, for example, F. Klein, who considered the genesis of the method of reasoning by recurrence to be "genuinely intuitive." Dedekind on the other hand, sympathizes with Frege's conviction when he states in the preface of his publication *Was sind und was sollen die Zahlen?* (What Are, and What Purpose Serve, the Numbers?) that the reader will be "frightened by the long series of simple inferences corresponding to our step-by-step understanding, by the matter-of-fact dissection of the chains of reasoning on which the laws of numbers depend, and will

[2] *Die Grundlagen der Arithmetik* (*The Foundations of Arithmetic*) V.

8. THE PRINCIPLE OF COMPLETE INDUCTION

become impatient at being compelled to follow proofs of truths which to his supposed inner consciousness seem at once evident and certain. On the contrary, it is precisely in this possibility of reducing such truths to others more simple, no matter how long and apparently artificial the series of inferences, that I recognize a convincing proof that their possession or the belief in them is never given immediately by inner consciousness, but is always acquired only by a more or less complete repetition of the individual inferences."[3] In this publication he tried to establish that the "form of demonstration known by the name of complete induction is actually conclusive, and that the definitions by induction are also definite and consistent." Poincaré would not be convinced by this "proof"; he said: "One can easily pass from one enunciation to another and thus get the illusion of having demonstrated the legitimacy of reasoning by recurrence. However, one will always encounter an obstacle, one will always arrive at an undemonstrable axiom, which will be in reality only the proposition to be proved translated into another language."

We meet another interpretation in the case of Russell. "The use of mathematical induction in demonstrations," so says this author, "was, in the past, something of a mystery. There seemed no reasonable doubt that it was a valid method of proof, but no one quite knew why it was valid. Some believed it to be a case of induction, in the sense in which the word is used in logic. Poincaré considered it to be a principle of utmost importance, by means of which an infinite number of syllogisms could be condensed into one argument. We now know that all such views are mistaken, and that mathematical induction is a definition, not a principle. There are some numbers to which it can be applied, and there are others" (the infinite cardinal numbers of Cantor are referred to here) "to which it cannot be applied. We define the 'natural numbers' as those to which

---

[3] Translator's note: cf. *Essays on the Theory of Numbers*, p. 33f.

8. THE PRINCIPLE OF COMPLETE INDUCTION

proofs by mathematical induction can be applied ... It follows that such proofs can be applied to the natural numbers, not in virtue of any mysterious intuition or axiom or principle, but as a purely verbal proposition. If 'quadrupeds' are defined as animals with four legs, then it will follow that animals that have four legs are quadrupeds; and the case of numbers that obey mathematical induction is exactly similar."[3]

We have just seen that opinions are divided regarding mathematical induction. We will now try to form an opinion about this subject.[4] We visualize the nature of complete induction very clearly in the example

$$
\begin{array}{r|l}
1:3 = 0\cdot 3 & 333 \\
10 & \\
\hline
10 & \\
10 & \\
\end{array}
$$

What this division shows is the periodic return of the remainder; and we conclude from this, that it "goes on in this manner without end." We may be tempted to say that the computation *yields the result* that all (infinitely many) digits of the quotients are threes. But here we must be cautious. Does the computation actually yield this result? Strictly speaking no, for every computation breaks off after a finite number of steps. It could yield, at the most, that the first ten, the first twenty, the first hundred digits were threes but not that *all* digits were threes. On the other hand, it is certainly obvious from the first step of the computation that the remainder will repeat itself, and that therefore the numerals of the quotient will repeat periodically. Hence, shall we still say that the general proposition follows from the computation?

[3] *Introduction to Mathematical Philosophy*, Chap. 3.

[4] For the ideas developed here, as well as for others which are quoted explicitly in the epilogue, the author is indebted to Ludwig Wittgenstein.

8. The Principle of Complete Induction

In order to clarify this point, we wish to use a fictitious situation. Let us suppose that somewhere there is a tribe which has our decimal system and computes just as we do, but is not cognizant of non-terminating decimal fractions. Everyone in such a tribe would be accustomed to break off a division, let us say, after the fifth place, since this is enough for their practical purposes. Thus, they would carry out the division 1:3 only to five places; as yet, it would not have occurred to them to ask what numerals are *next*. Let us now assume that one day someone were to get the idea that the division could be carried out to any number of places and that the decimal fraction which thereby arose had an infinity of threes. What would his discovery consist in? When this becomes clear to us, we can expect to obtain a deeper insight into the nature of induction.

At first one may be tempted to say that the discovery consists in the fact that he was struck by something which up to now the others had paid no attention to, namely, the repetition of the remainder. However it would not be quite correct to say this. For, if an individual who still did not know about periodic division had been asked: "Is, in this division, the first remainder equal to the dividend?" he would naturally have said "yes"; hence his attention would have very definitely been attracted by it. Nevertheless, he need not have noticed the periodicity.

Perhaps one would now want to say that whoever discovers the periodicity *sees* division differently than one who does not know of it, for he reads into it an infinite possibility. However, this now sounds as if it depends on something psychological, on a kind of vision. *In reality the discovery of the periodicity is the construction of a new calculus.* It must now be clear to us that the division $1 : 3 = 0.\overline{3}$ is not a computation of the same kind as $1 : 2 = 0.5$, rather the division

$$1 : 2 = 0.5 \text{ corresponds to } 1 : 3 = 0.\overline{3}$$
$$0 \qquad\qquad\qquad\qquad\qquad 1$$

8. THE PRINCIPLE OF COMPLETE INDUCTION

namely, to a *finite* computation. One could also say that 0.3 is not the result of division in the same sense as 0.5. The latter symbol was known to us before the division 1 : 2. What, however, does 0.3 mean detached from the periodic division? The statement that the division yields the quotient 0.3 *simply means* that the first digit of the quotient is 3 and the first remainder is equal to the dividend; and this statement exactly describes the periodicity of the computation. The distinction between this and an ordinary division can be expressed by stressing the repetition of the remainder, for example, one kind may be written

$$1 : 3 = 0.3, \text{ the other kind } \underline{1} : 3 = 0.3$$
$$1 \qquad\qquad\qquad\qquad \underline{1}$$

"But this is only a formality," the reader will say. "The character of the computation is not changed by merely drawing a dash." However, the dash refers to the *law* in the division; thereby we have actually gone over to a new kind of calculation. Whoever calls attention to the periodicity introduces, by this very same procedure, a new symbol. This means that the way by which he calls attention to the periodicity gives the new symbol.

If I break off the division after some steps and say "only threes will follow from now on"—do I thereby overleap the abyss between the finite and the infinite? Must I have recourse to a mysterious principle which contains infinitely many inferences in one act? That is out of the question. The illusion which we succumb to here is that we seem to speak of an *infinite extension* (the infinitely many numerals which are not written down), whereas, in the discovery of the periodicity, we have instead discovered a *law* which yields numerals. Strictly speaking, all that can be said is that 0.333 ... is not an abbreviation for a number which cannot be written down completely only due to the lack of ink and paper; instead it is a new symbol which has its own (and indeed finite) rules of calcula-

8. The Principle of Complete Induction

tion. In other words, the dots do not, in a shadowy fashion, stand for the numerals which are not written down, but they are themselves a full-valued symbol in our calculus.

Now the situation begins to clear up. The "jump from the finite to the infinite" is actually the transition to a new calculus which is not a logical consequence of the old. This new calculus cannot by any means be deduced from the old even though it leans upon the latter in a definite manner. As long as only ordinary division was known to us, we could not deduce from the division 1 : 3 that the numeral 3 repeats periodically. In making the discovery of the periodicity, we have discovered a new calculus.

After these remarks, the nature of induction should be more intelligible to us. It must strike us that a proof by induction possesses a structure entirely different from what is called a "proof" in algebra. A proof, formally considered, is indeed a sequence of formulae which starts from known formulae and ends with the one to be demonstrated, where the transition from one formula to the next takes place according to fixed rules. If this is the case, then we say that the last formula of the chain is demonstrated. However, if we consider a proof by induction, an entirely different situation is met, for the proof does not lead at all to the formula to be demonstrated. As an example, we will recall the proof of the associative law of addition. We started from the definition

(D) $\qquad a + (b + 1) = (a + b) + 1$

and the assumption that the proposition was already valid for a definite c:

(A) $\qquad a + (b + c) = (a + b) + c.$

We then proved that the proposition was also valid for the number $c + 1$:

$$
\begin{aligned}
a + (b + [c + 1]) &= a + ([b + c] + 1) & &\text{(by D)}\\
&= (a + [b + c]) + 1 & &\text{(by D)}\\
&= ([a + b] + c) + 1 & &\text{(by A)}\\
&= (a + b) + (c + 1) & &\text{(by D)}
\end{aligned}
$$

8. The Principle of Complete Induction

It is obvious that the theorem to be demonstrated never occurs in this proof. Furthermore, we do not say that the associative law has been *calculated* in the sense in which perhaps other equations are calculated with the help of basic laws. Instead, the usual view is that a special inference is added to this chain of equations, which says: *therefore* the proposition is valid for all numbers. By this inference we seem to overleap the abyss between the finite and the infinite. It is remarkable how a biased manner of expression can so disguise the problem that the way back to the correct view can be found only with difficulty.

Above all we must ask: is the induction only the *symptom* of the fact that the proposition is valid for all numbers? Or do the words "the proposition is valid for all numbers" mean nothing more than: "it is true of 1; and if it is true of c, it is also true of $c + 1$?" Usually we think of the proposition as merely stating that something is true for all numbers, and the proof by induction is only one of the ways which lead us to the recognition of its truth. Hence, a distinction is drawn between the proposition—which has a meaning when taken by itself—and the proof, which, as it were, only puts us on the way to it. However, what shall the general proposition mean if I disregard the proof by induction? Perhaps, that it is true of the number 1, of the number 2, etc., to infinity? However, this does not explain the meaning of the proposition, namely, it cannot answer the question: how should this proposition be used? What do we regard as a criterion of its truth? We certainly could not run through all numbers and make infinitely many trials; not merely because we do not have enough ink and paper but because it means nothing and is logically impossible. Actually the proof by induction is the *only* criterion we have. But then, it is only the proof itself which tells us the meaning of the proposition. To generalize: if we wish to know what a proposition means, we can always ask, "How do

8. THE PRINCIPLE OF COMPLETE INDUCTION

I know it?" Its meaning is determined by the answer to this question.[5]

This remark contains the key to understanding induction. It is quite natural to say that the proof shows that the proposition is valid for all numbers; however, we must then be mindful that the sense of the word "all" is determined only by the proof. And this sense is different from the one in the example: "All seats in this room are made of wood." For to negate the latter statement means that "there is at least one seat in this room which is not made of wood"; whereas, to negate the statement "A is valid for all natural numbers," only means that one of the equations in the proof of A is false, but not that there is a number for which A is not valid (unless we wished to define the meaning of this statement by this criterion).—— This is also substantiated by the following argument: to negate a generic formula, for example, the formula $(a + b)^2 = a^2 + ab + b^2$, simply means that it is not this formula which is valid but . . ., and now the correct formula is stated. Hence the contradiction only succeeds in bringing one generic formula in contrast to another generic formula, but not in forming an existence statement. It is very similar to saying: not grey but yellow, not 2 but 3, where the negation of one statement is only the preparation for another. The situation is entirely different from that in the example: not all old men have grey hair, which

---

[5] These remarks can also be applied to certain problems which are considerably discussed today, for instance, the legitimacy of existence proofs. To the intuitionists an existence proof is valid only if it is constructive, i.e., if it indicates a procedure given for constructing the object under consideration in finitely many steps. The intuitionists repudiate all other existence proofs as "meaningless"; while the formalist also permits non-constructive proofs. If our remark is correct, it shows the futility of this entire dispute. For, the word "existence" of itself does not have a clear-cut meaning; it acquires one only by proof. Now, if the proof is in one case constructive, in another case non-constructive, it just means that the existence statement has a *different meaning*.

8. THE PRINCIPLE OF COMPLETE INDUCTION

merely says that there are old men whose hair is not grey. In other words, the generic formulae of mathematics and the existence statements do not belong to the same logical system at all. Brouwer understood this correctly when he noted that the incorrectness of a general statement about numbers by no means implies the existence of a counter-example.

Now the accomplishment of the induction is clear. It is not a conclusion which takes us to infinity. The proposition $a + b = b + a$ is not an abbreviation for infinitely many individual equations which cannot be all written down due to our human weakness—no more than $0.333\ldots$ is an abbreviation for an infinitely long decimal fraction—and the inductive proof is not an abbreviation for infinitely many individual syllogisms. To comprehend induction in this way seems only to obstruct the path to its understanding.

When we set up the formulae
$$a + b = b + a$$
$$a + (b + c) = (a + b) + c$$
etc.

we are actually beginning an entirely new calculus, a calculus which cannot be deduced from the computations of elementary arithmetic. This is what is correct in Poincaré's remark that the principle of induction is not demonstrable in a logical way. However, neither does it represent, as he thinks, a synthetic a priori judgment. Instead of being a truth it is merely a convention whereby, if the formula $f(x)$ is true for $x = 1$, and $f(c + 1)$ follows from $f(c)$, we say that "the formula $f(x)$ is proved for all natural numbers."

"However," one will ask, "is it really true that this is only a convention? Do we not see that the proposition is actually correct for any individual number which we may single out?" It could appear paradoxical that the associative law of addition follows from a mere definition (the formula D). However, we should not forget that formula D is not a definition in the sense of school logic, namely a substitution rule, but *a direction for*

8. THE PRINCIPLE OF COMPLETE INDUCTION

the formation of definitions. Hence the universality already lies in the definition, and from there it is transferred to the inductive proof. The induction proof can also be thought of as a direction for the formation of proofs for individual numerical equations, as the general term of a series of proofs. Indeed, the induction proof could very well be written in the form of a series of equations with individual numbers, as a part of a series with an "etc.," and it would thereby lose none of its power. This manner of writing also shows much more clearly that the generic formula does not *follow* from the induction proof—in the formula, of course, letters occur, in the proof, however, only numerals. However, a new convention can be agreed upon which allows the transition to the generic formula.

Thereby a riddle is cleared up which has deeply disturbed the scholar. How does it happen that the result of a particular computation can be *predicted* without carrying it out? For example, why do we know that 63×289 will give the same result as 289×63? This uneasiness arises whenever we do not see the dependence between the general prediction and the particular computation. On this account, one was always inclined to think of the formula $a \cdot b = b \cdot a$ as a condensation of all individual computations. Actually this is not the meaning of these formulae. Instead, the commutative law can be compared to an arrow which points along the number series to infinity. This is not the same as saying that the law contains infinitely many individual propositions. The difference is about the same as in the sentences: the reflector beams to infinity, and: it lights up the infinite.

Thus we can say that by agreeing on that convention—namely, to set up formulae which meet the demands of inductive proofs—we adapt the calculus with letters to the calculus with numbers. This means that we bring this calculus in harmony with the calculus of natural numbers as established by the recursive definition of addition, etc.

9. Present Status of the Investigation of the Foundations

A. *Formalism*

We now resume the account of the further progress made in investigating the foundations of mathematics and first turn to formalism. The considerations which led to the development of this way of thinking are known. It was primarily the problem of consistency which forced the mathematician to a rigorous formulation of the mathematical propositions and methods for drawing conclusions. Has the attempt to prove consistency been successful?

Up to a short time ago it looked as if Hilbert's methods of attack would lead to the desired end; indeed, in the case of certain parts of mathematics, for example, the elementary theory of numbers, the consistency seemed to have been already demonstrated, and actually by finite-intuitive methods. In other words, it appeared that the consistency of the theory of numbers including logic could be proved through a metamathematical investigation which used only a part of the assumptions of the theory of numbers and logic. Meanwhile, however, the situation has changed essentially. An investigation of Gödel[1] dis-

---

[1] *Über formal unentscheidbare Sätze*, Monatshefte für Mathematik und Physik (*About Formally Undecidable Propositions*, Monthly for Mathematics and Physics), 1931.

9. Present Status of the Investigation of the Foundations

closed the surprising result that *the consistency of a logico-mathematical system can never be demonstrated by the methods of this system*; instead, the proof of consistency must involve essentially new methods which are not expressible within the system itself. Gödel's insight can be formulated more exactly as follows: If the axioms of logical calculus are added to the axioms of Peano and the resulting system is called P, then the consistency of P cannot be demonstrated by a proof which can be formulated in P, provided that P is consistent. (The latter statement is essential. For if P were self-contradictory, then *any arbitrary* formula could be deduced from the axioms of P; in particular, the formula which says, referring to its content, that P is consistent.)

At first this situation—the impossibility of formulating the consistency of P within P—will seem to exist because the system under consideration is devoid of expressions, and therefore we should strive to extend it in order to attain this end. However this would not help. For Gödel has established the very general proposition: Whatever formal system we may consider that includes the theory of natural numbers and logical calculus (to be more precise, that arises from P by the addition of a recursively defined class of axioms) an argument will always appear in the proof of consistency which *cannot* be expressed within this system.

However, this important result is only the consequence of another proposition of Gödel which is much more fundamental. It can be roughly expressed as follows: *Every arithmetic is incomplete*; namely, *in each of the formal systems mentioned before there are undecidable arithmetical propositions and, in each of these systems, arithmetical concepts can be found which are not definable within this system.* (For example, in every formal system S a real number can be constructed which cannot be defined in S.) This statement should not be interpreted as proving that there are mathematical problems which are definitely unsolvable; it only means that the concept "solvable" or "decidable" always refers to a *definite* formal system. If a proposition is undecidable in a particular system, the possibility of constructing a more comprehensive system in which the proposition can be decided is always feasible. However, there is no system in which *all* arithmetical propositions could be decided or in which *all* arithmetical concepts could be defined. When Brouwer thinks of mathematics as essentially a living activity of thought, a series of meaningful constructive steps and *therefore* not contained in a rigid system of formulae, he is on the right track. "All mathematics can be formalized; however, mathematics can *never* be exhausted in *any one* system but requires an infinite sequence of discourses which get progressively more comprehensive."[2]

This non-closure of mathematics is not due to a defect in our human makeup, but lies in the nature of the subject. We had previously visualized mathematics as a system all of whose propositions are necessary consequences of a few assumptions, and in which every problem could be solved by a finite number of operations. The structure of mathematics is not properly rendered by this picture. Actually mathematics is a collection of innumerably many coexisting systems which are mutually closed by the rules of logic, and each of which contains problems not decidable within the system itself. *One* of these problems is that of consistency; in other words, the statement that the system S is consistent is undecidable in S.

However, this does not mean that the end sought by Hilbert is unattainable. In any case we will have to alter the starting point, namely, the restriction to a primitive part of arithmetic and logic. It is uncertain today whether the consistency of classical mathematics (arithmetic, algebra, analysis, function theory) can then be proved. A work of Gentzen

[2] Carnap: *Die Antinomien und die Unvollständigkeit der Mathematik*, Monatshefte für Mathematik und Physik (*The Antinomies and the Incompleteness of Mathematics*, Monthly for Mathematics and Physics), 1934.

9. Present Status of the Investigation of the Foundations

which appeared recently, has made an important contribution in this direction. In this work the consistency of the whole of arithmetic is actually proved on the basis of a part of arithmetic which does not include the law of the excluded middle and certain transfinite methods.[3]

However, even if the desired end could be reached in this way, a second question would raise itself to which we must attach the greatest importance. Can arithmetic be founded at all by investigations of this sort? It will become clear that this question is quite justified as soon as arithmetic is compared to geometry. The problem of a "foundation for geometry" implies two distinct problems. First, a set of propositions must be selected which form a logical basis for the geometry under consideration, and the independence, completeness and consistency of these basic propositions must be proved. Secondly, since distinct geometries are conceivable, the question arises how these systems can be applied to our world of experience—the topology of rigid solids, the behavior of light rays, of inertial paths, etc. This second question represents a problem of natural science which cannot be solved by mathematical thinking alone.

On applying these considerations to arithmetic, we recognize that, at best, axiomatical investigations will settle the first problem. There remains the question to what extent the axioms of arithmetic hold for "actual numbers," that is, for those meanings which we attach to the numerals. This question represents a new and very profound problem, which should not be neglected in the formal construction.

In order to make our ideas more precise, let us consider Peano's five axioms. Peano assumed that we already knew the meaning of the words "zero," "number," "successor," and that

---

[3] G. Gentzen, *Die Widerspruchsfreiheit der reinen Zahlentheorie*, Mathematische Annalen (*The Consistency of the Pure Theory of Numbers*, Mathematical Annals), 112.

103

9. Present Status of the Investigation of the Foundations

the axioms are manifestations of a truth. The formalist does not share this view. As far as he is concerned, the axioms are meaningless combinations of symbols whose structure alone interests him. The symbols "zero," "number," "successor" can then be interpreted in an infinity of ways, and each of these interpretations would impart another meaning to arithmetic. Russell gives some instructive examples.

1. Let "0" stand for 100 and let "number" be taken to mean the numbers from 100 onward. All our basic propositions are satisfied. Even the fourth is valid; for, though 100 is the successor of 99, 99 is not a "number" in the sense which we are now giving to the word "number." Obviously, instead of 100 any arbitrary number could be used in this example.

2. Let "0" have its usual meaning, but let "number" stand for an "even number," as this term is ordinarily used, and let "successor" of a number indicate the number which results from adding 2 to it. Then "1" will stand for the number two, "2" will stand for the number four, etc. The number series will now read

$$0, 2, 4, 6, 8, \ldots$$

All five assumptions of Peano are again satisfied.

3. Let "0" stand for the number one, let "number" stand for the sequence

$$1, \tfrac{1}{2}, \tfrac{1}{4}, \tfrac{1}{8}, \tfrac{1}{16}, \ldots,$$

and let "successor" mean "half of." All five axioms of Peano also hold for this sequence.[4]

These examples show that the axioms do not characterize the concept of number series. Instead they do characterize a much more general concept, namely, that of *progression*. Every progression satisfies Peano's five axioms. A progression need not be composed of numbers; it can be composed, just as well, of a range of points on a straight line, or of moments of time, etc. Now, since there are infinitely many progressions, Peano's

[4] *Introduction to Mathematical Philosophy*, Chap. 1.

9. Present Status of the Investigation of the Foundations

system of axioms can be interpreted in an infinity of distinct ways. Perhaps the reader thinks that we should be resigned to the view that by a number we mean any entity whatever which satisfies Peano's axioms, and we must refrain from investigating their meaning just as we refrained from defining the basic concepts in geometry. Though such a point of view is justified for some purposes, in general it cannot be accepted, for we also use numbers in mathematics as a means of communication. For example, in saying, "There are only 5 regular solids," I want to assert a true proposition unambiguously. However, it is impossible to attach a definite meaning to this statement if "5" is a mere counter which can be interpreted in the most varied ways. On adopting such a standpoint, we would no longer be able to distinguish the two communications. "There are 5 regular solids" and "There are 105 regular solids." Thus, the consideration of the simplest example shows that if we restrict ourselves to this standpoint the very sense of the number statement is allowed to escape.

Does this situation arise because the axioms of Peano are still too indefinite to determine the number series and is it perhaps possible to narrow down this indefiniteness by the addition of further axioms which will yield a complete characterization of the sequence of cardinal numbers? It is now extremely significant that Skolem has thwarted every hope of this kind. That is, he proved a general proposition which says that it is impossible to characterize the number series by finitely many axioms.[5] For, every statement which is valid in the arithmetic of natural numbers is also valid for structures of another kind, so that it is impossible to distinguish the number series by any inner properties from sequences of another kind.

---

[5] *On the Impossibility of Completely Characterizing the Number Series by Means of a Finite System of Axioms* (title translated from the Norwegian), Norsk. Math. Forenings Skrifter, Ser. II. 1933.

9. Present Status of the Investigation of the Foundations

The possibility of reinterpreting the concept of integer, with so many propositions required to be true that one might be tempted to believe they could hold only for integers, may be demonstrated by the following striking and nontrivial example.[6] The integers are

1. linearly ordered;

2. they reproduce themselves by addition, subtraction and multiplication. This addition is commutative and associative, and the same is true of multiplication; furthermore, both operations satisfy the distributive law. There are two numbers which are next to one another, namely 0 and 1, such that for any arbitrary number a: $a + 0 = a$ and $a \cdot 1 = a$;

3. if the concepts of divisibility, unity, relatively prime etc., are introduced in the usual manner, the following proposition is valid: if a and b are relatively prime, there are two numbers x and y such that $ax - by = 1$.

Let us now consider all polynomials of the form

$$a_n t^n + a_{n-1} t^{n-1} + \ldots + a_1 t + a_0,$$

where a_0 is an integer, while $a_1, a_2, \ldots a_n$ are arbitrary rational numbers. Let us think of these polynomials as ordered lexicographically. (This means that if two polynomials have the coefficients $a_n, a_{n-1} \ldots a_0$ and $b_n, b_{n-1}, \ldots b_0$, the first is smaller than the second if $a_n < b_n$, or, in case $a_n = b_n$, $a_{n-1} > b_{n-1}$ etc.) This polynomial domain satisfies all properties which we have cited in 1 to 3—even though they seemed originally to characterize the concept of integer. Skolem's proposition now says that such a new interpretation will always be possible no matter how many properties are used to set up the concept of natural number.

---

[6] Cf. Skolem, *On some Questions Regarding the Foundations Mathematics* (title translated from the Norwegian), Skrifter nors Vid.-Akad., Oslo, I. Mat. Nat. Kl. 1929.

9. Present Status of the Investigation of the Foundations

B. *The Logical School*

These results show, at any rate, that the last word regarding the nature of arithmetic is not to be found in formalism. And the reason for this is also not difficult to discover: this way of thinking is one-sidedly preoccupied with the structure of the mathematical statements and, on the other hand, fails to investigate the mathematical concepts. The logical school, on the contrary, has concerned itself precisely with the analysis of the concepts. The following assertion due to Frege illustrates this standpoint very expressively: "Strictly speaking, it is really scandalous that science has not yet clarified the nature of number. It might be excusable that there is still no generally accepted definition of number, if at least there were general agreement on the matter itself. However, science has not even decided on whether number is an assemblage of things, or a figure drawn on the blackboard by the hand of man; whether it is something psychical, about whose generation psychology must give information, or whether it is a logical structure; whether it is created and can vanish, or whether it is eternal. It is not known whether the propositions of arithmetic deal with those structures composed of calcium carbonate or with non-physical entities. There is as little agreement in this matter as there is regarding the meaning of the word 'equal' and the equality sign. Therefore, science does not know the thought content which is attached to its propositions; it does not know what it deals with; it is completely in the dark regarding their proper nature. Isn't this scandalous?"[7]

Frege was one of the first to attack the problem of defining number. He defines the concept of number in two steps. First he explains when two sets are numerically equivalent (gleich-

---

[7] *Über die Zahlen des Herrn H. Schubert* (On the Numbers of Mr. H. Schubert).

zahlig); then he formulates a definition of the concept "number" (Anzahl) in terms of the concept of numerical equivalence. According to Frege, two sets are numerically equivalent if to every element of one set there corresponds an element of the second, and to every element of the second set, an element of the first. In other words, two sets are numerically equivalent if there is a one-to-one relation which associates the sets to one another. By the "number of a set" he understands the set of all sets numerically equivalent to it. Accordingly, the number 5 is defined as the class of all classes in the world with five elements, in other words, it is the totality of all those classes which can be uniquely related, for example, to the class consisting of the fingers of my left hand.

Perhaps the reader will find this definition somewhat odd; he will say that this is not what he means by the number 5. However this would still not be a sufficient reason for rejecting this definition, provided that in other respects it fulfilled the requirements imposed on a good definition, namely, that it actually render the generally recognized properties of the concept. But does it do this? Let us now take up this question.

We have scruples even about the first part of the definition. Two sets are said to be numerically equivalent if they are related to one another by a single-valued relation. How then shall we establish that two sets are numerically equivalent? Evidently we must exhibit such a relation. If I wish to establish, let us say, that I have as many spoons as cups, according to this precept I must find a single-valued relation which associates each spoon to a cup. For example, such a relation could be: every spoon lies in a cup and no cup is left empty. But, suppose the spoons are in a box and the cups in another? Is there a relation which associates them to one another in this case too? We could say that "If there isn't one, it is at any rate very easy to set one up; I only have to distribute the spoons among the cups." We merely remark that this would mean that the relation at least did not exist beforehand. Consequently, we would have

9. Present Status of the Investigation of the Foundations

to say that as long as the spoons were not in the cups the two sets were not numerically equivalent, which is an interpretation that does not correspond to the sense in which the word "numerically equivalent" is used.

One will reply that this was not the intention of the explanation; it does not depend on whether I actually place the spoons in the cups but whether I *can* place them in the cups. Very well! But what does the expression "I can" mean here? Is it that I have to be physically able to distribute the spoons among the cups? This would be entirely uninteresting. Obviously, what we wish to say is that I can distribute the spoons among the cups because there are just as many samples of both sorts. That is, in order to recognize whether the correspondence is possible, I must already know that the sets are numerically equivalent. Therefore the numerical equivalence is not determined by the correspondence, but the numerical equivalence makes the correspondence possible.

It is now obvious how the deception comes about. The statement: "The two sets *can* be associated to one another," is being reinterpreted into the statement which is entirely distinct from it: "The two sets *are* associated to one another," which means, "There is actually a relation which permits such a correspondence." Only experience can tell us whether or not such a relation exists; and there is nothing to prove that it always exists.

To recapitulate we can say that the proposed definition gives a sufficient, but not a necessary condition for numerical equivalence. Therefore it restricts the meaning of the expression "numerically equivalent" too narrowly.

What is the meaning of the expression "numerically equivalent"? If we wish to obtain a clearer standpoint regarding this question, we will first of all have to draw a distinction. Hitherto, it is true, we have spoken of classes, but have failed to explain the meaning of this expression. The word "class" is used in two ways: first, to designate those things which are enumerated in a list (school class); secondly, as having the same meaning as the word "concept." ("The whale belongs to the class of mammals" means "The whale falls under the concept of 'mammal'.") These two meanings are completely distinct, for in the first case, it is a tautology to say that a belongs to the class k (for this follows from the definition of k) while, in the second case, it is an empirical cognition.

If we now ask when are two classes numerically equivalent, our answer will depend on what we understand by a class. If we are dealing with two concepts, it can be empirically established whether the extensions (Umfänge) can be associated to one another; even if one set should contain more elements than the other, the statement: "They can be associated to one another," is only false, not self-contradictory. However, if we consider the classes as two lists, the situation is different; in this case, neither the statement: "They can be associated to one another," nor: "They cannot be associated to one another," is an empirical statement. Two examples may illustrate this point.

1. "Are there as many persons in this room as in the next room?" A primitive way of deciding this question would be the following: from every person in one room, draw a string to a person in the other room such that a string actually leads to every person and no string is split. The spanning of the string is here an *experiment*. Before the experiment is carried out we could have said: the two sets can be associated to one another, as well as, they cannot be associated to one another.

2. "Are 3×4 cups numerically equivalent to 12 spoons?" Here also the decision could consist in the drawing of lines. Thus we could represent the cups by a series of points, similarly for the spoons, and then draw a line from each point of one series to a point of the second series. The drawing of the lines is in this case not an experiment but a procedure in calculus, since this is the way we calculate whether the two sets are numerically equivalent or not. The proposition: "The

9. Present Status of the Investigation of the Foundations

are associated to one another," is now a proposition of mathematics.

Let us now return to our starting point. According to Frege's definition two sets are numerically equivalent if there is a one-to-one relation which associates them to one another. We first objected to saying: if such a relation is set up, the sets are numerically equivalent; while if it is not set up, they are not numerically equivalent. True, such an explanation has defined something, but not the concept "numerically equivalent." This definition has then been improved by explaining that two sets are numerically equivalent if they *can* be associated to one another by a one-to-one relation, that is, if it makes sense to say that they are associated to one another by a one-to-one relation. However, this is again not quite correct. For, if the two sets are given by their properties, it is *always* meaningful to maintain that they are associated;[8] however, if two lists are submitted, for example, the lists (a, b, c) and (a′, b′), it is a contradiction to state that they are associated.

Actually there are various criteria used for the expression "numerically equivalent." Frege stresses only *one* of these and sets it up as a paradigm. Let us consider some examples.

1. If three spoons and three cups are on the table, we establish at a glance that they are numerically equivalent without having to relate them to one another term by term.

2. If the number is not surveyable, but the things are arranged in a surveyable array as, for example, the collections of points

---

[8] The reader should notice that the meaning of the statement: "The two sets can be associated to one another," changes with the criterion used to recognize the possibility of the correspondence, that is, according to whether we say that the two sets are *numerically equivalent*, or that it shall have a *sense* to speak of a correspondence.

9. Present Status of the Investigation of the Foundations

```
  • • •         •
  • • •       •   •
  • • •       •   •
                •
```

the numerical equivalence is again obvious.

3. The situation is somewhat different when we determine, let us say, that two pentagons have the same number of diagonals. In this case we can no longer directly apprehend the grouping of the diagonals; rather, it is a geometrical proposition that the two classes are numerically equivalent.

4. Two sets are numerically equivalent if there exists a one-to-one relation between their elements.

5. However, the normal criterion for the equality of numbers is counting (which must not be regarded as a mapping of two sets by a relation).

Frege's definition does not render these dissimilar and flexible uses; it recognizes only *one* pattern in which it would like to press all cases.

It is easy to show that this view leads to unusual consequences. According to Frege, two sets must either be numerically equivalent or not; and, what is more, from a purely logical basis. Let us assume that someone glanced at the sky bright with stars and was then asked how many stars he had seen—can the answer be: "I do not know exactly how many stars were there, but it must have been a definite number"? How is this answer to be distinguished from the following: "I have seen many stars"? (It should not be forgotten that we are asking about the number of stars actually *seen*, not the number of physically existing stars, which can naturally be counted on a photographic plate.) Perhaps one says: "I have forgotten the exact appearance of the sky, but at the moment when I was looking at it I must have seen a definite number of stars." This means that if I could only take another look

9. Present Status of the Investigation of the Foundations

at it, as it was, I could readily tell how many stars were there. However, the irony of the situation is that there can be no going back. The event is over; I cannot hold on to it in order to contemplate it afterwards at leisure. There is just no method for determining the number; and consequently the numerical statement loses its sense.

However, we could also look at this matter in another way. Since a small number of stars—let us say, up to 5—can still be apprehended clearly, and not until then does their number begin to grow hazy, we could say: Let us apply here a new number series whose terms are "1, 2, 3, 4, 5, many"—a series which is actually used by many primitive tribes. The reader will perhaps have misgivings at calling this a *number series;* he will regard our proposition more as an evasion of a difficulty. We only leave it to the consideration of such a reader that this number series is by no means incomplete and we do not have one which is any more complete; instead we are only in possession of a different and more complicated one, besides which this primitive series has the right to stand. We can also add and multiply in this series, and all this can be done completely rigorously; but, it is adjusted to other purposes. Let us imagine there were no durable things in this world; instead, that everything were to float past like the images of falling rain drops. Perhaps this number series would then appear just as natural and appropriate to us as the one that we are used to. As another example: let us suppose that while we were counting objects some were to disappear or new ones to arise without our noticing it, so that another number would always be discovered. Such experiences could direct our formation of concepts along entirely different paths. Perhaps indefinite numerals such as "few," "many," "very many" — eventually with finer gradations — would replace our present series of numerals. On heeding such possibilities, we free ourselves of the prejudice that our number series is a distinguished structure, as if it

9. Present Status of the Investigation of the Foundations

were a God-given gift to us, which must be essentially the cornerstone of what we call arithmetic.

Frege's definition pays no attention to all this. According to him two sets are numerically equivalent or not by virtue of a purely logical law, whether we can now confirm it or not. It was argued earlier exactly in this way that the simultaneity of two events depended only on the events themselves, and was entirely independent of the observations by which we recognize the simultaneity. However, this is not the case; instead the meaning of a statement exhausts itself in the way it is verified. If we wish to know the meaning of the expression "numerically equivalent," we must be mindful of the methods by which the numerical equivalence is actually established. Now these proceedings are very complicated and not to be brought into a scheme as simple as Frege thought it to be.

We will devote only a few words to the second part of Frege's definition. This says that a number shall be a class of classes. According to this definition the statement: "Three apples lie on the table," would mean that the number 3 is predicated of the concept "apple which lies on the table"; in other words, the class of the apples lying on the table is an element of the class 3. This interpretation, which even Russell thinks of as somewhat paradoxical, has *one* great merit: it places in evidence from the beginning that number states something about the concept not about the counted things themselves. Therefore it avoids certain errors which Frege has justly brought into prominence. However, will this interpretation actually take into account the uses of the numerals? Let us imagine a language which would be used exclusively to give commands, whereas it could not be used at all to formulate statements. Now what does it mean in this language to give the command: "3 apples!"? Does this assign another meaning to the numeral "3"? Certainly not! Yet this command can no longer be properly explained by Frege's scheme; it does not say that the class of apples to be called for is an element of the class 3.

9. Present Status of the Investigation of the Foundations

For this is a statement, and therefore our language is not aware of this interpretation. The command cannot be expressed at all in the form of subject and predicate. *Hence Frege's definition unnecessarily restricts the concept of number to the subject-predicate form of our propositions.* Actually the meaning of the word "3" results from the way in which it is applied, and we will speak about this later.

A second difficulty is illustrated most effectively by an argument of Russell. Let us assume, says Russell, that there were exactly 9 individuals in the world. Then we could construct the cardinal numbers from 0 up to 9, but 10, which would be defined as $9 + 1$, would be the null class. Consequently 10 and all subsequent natural numbers would be identical, since they would all be 0. In order to prevent such an arithmetical catastrophe a special axiom must be introduced, the axiom of infinity. This says that there is a type composed of infinitely many individuals. The axiom just mentioned represents a statement about the world, and the structure of the whole of arithmetic now depends essentially on the truth of this axiom. Therefore everyone will be eager to discover whether this axiom is true. Regarding this point we must reply that nothing is known; in fact, we have not even the slightest reason for believing in the truth of this axiom. But still worse, we can never reach a decision regarding the truth of this axiom, since its nature is such that it escapes any examination. Thereupon we must take the stand that no meaning can be assigned to this assumption, for what shall a hypothesis mean which is eternally uncertain? Accordingly, it would be possible to base arithmetic on an assumption about the world which cannot be verified, on a meaningless hypothesis. If this interpretation were to be submitted to impartial minds, who would be convinced? Who will still believe that the theory of natural numbers will thereby be more deeply founded? Instead we will feel justified in judging that there must be something definitely out of order with a theory which only makes the simple and clear concept

115

9. PRESENT STATUS OF THE INVESTIGATION OF THE FOUNDATIONS

of number obscure and problematical, and that it is of no value as an explanation of the concept of number.

It also does not help to avoid the difficulty by carrying along the axiom of infinity as a condition in the mathematical propositions. For we do not thereby gain the mathematics which actually lies before us. In mathematics we say, for example: the set of fractions is everywhere dense, but not: the set of fractions is everywhere dense *if* the axiom of infinity holds. This is an artificial reinterpretation, only devised to maintain the theory that the numbers are constructed out of actual classes in the world.

We will now summarize our scruples. In the case of formalism, it appears to us that the interpretation of numbers as meaningless symbols is not compatible with the applications which these symbols already find within mathematics. In the case of the Frege-Russell theory, we recognize that it endeavors to give a meaning to the number symbols. In so doing, however, it ties this interpretation to actual classes in the world, brings in an empirical element which is accidental to arithmetic and thereby destroys the apriority which is characteristic of mathematics.

C. *Outlook*

Hence, we have still not disposed of the question: What is a number? None of the directions previously developed have given us a clear answer. It could almost look as if the simple concept of number contained a mystery which resisted its being grasped by the intellect. However, as often in such cases, we should first consider whether a false formulation is not at the source of this failure. The question: "What is a number?" is related in many respects to the question of Augustine: "What is time?" We all understand this word and know how to use it in daily life. However, if asked what time is, we get into

9. Present Status of the Investigation of the Foundations

trouble and would like to use the words of Augustine: "If I am not asked, I know it; if I am asked, I do not know it."

In order to clear up this riddle we only have to consider what we would do if we wanted to explain to someone the meaning of this word. We would show him by various examples how the word is used; for instance, we could cite the cases: "I have no time now," "Now is not the time for this," "My time is up, I must go," "About what time will you come?" "The time has been passing very fast," "I had to wait a long time," "Once upon a time there lived . . ." Hence to answer Augustine's question we give types of application. We combine this word with other words; we fit it into different syntactical combinations; we follow, as it were, all the paths which the usage of language has prepared for this word, and in this way explain its meaning. In fact, whoever is able to understand the word "time" in the various examples and to apply it, knows just "what time is" and no formulation can give him a better understanding of it. The question: "What is time?" leads us astray, since it causes us to seek an answer of the form "Time is . . .," and there is no such answer.

If the question: "What are numbers?" is approached from this point of view we will no longer succumb to its suggestive spell. Instead of setting up the nature of number in a formula we will describe the uses of the word "number" and of the numerals.

How, then, should we explain the numerals? Is there perhaps a demonstrative definition? For the beginning of the series there certainly is. For instance, we can point to some intuitively evident group of things or circumscribe them with a motion and say: 2 nuts, 3 apples, etc.; and this explanation is not bad. True enough, it breaks down in the case of larger numbers; here we note the existence of a certain distinction which could be designated by the words "visual number," "counted number." The fact that in Greek the first four numerals are declined but not the others indicates that a

117

9. Present Status of the Investigation of the Foundations

distinction in the interpretation of number had been perceived earlier.

Normally the usage of numerals is tied to counting. We need not depict how the counting is done. Instead, we will anticipate an objection which may already have been stirred up, namely, that this is a method by which children learn the numerals, but not a scientific explanation of the number concept. This must be exact and complete. However, the meaning of the numerals results from their applications, and there are many of these. If a command such as "6 apples!" prompts a child to count from 1 to 6 and to lay down an apple with each word — isn't he now clear about the sense of the word "6"? Does he still only have a crude or incidental idea of the number 6? Must the logician first find out its true sense? All of us have certainly learned the numerals in our childhood in this way and understood them ever since; and we know of no better explanation. Why therefore should we be ashamed of it? Actually the process by which children learn numbers in school — with strokes, beads, etc. — is perfectly correct.

Consequently we must first be clear regarding our objective. We do not seek a definition of the number concept, but a clarification of the grammar of the numerals and the word "number." The real reason why we strive for a definition is simply because we think that arithmetic can thereby be more deeply founded. The propositions of arithmetic should be demonstrated as purely logical truths. May we take the liberty of pointing out at least our position regarding this question.

In our opinion mathematics does not consist of tautologies. It is not a branch of logic, but completely autonomous and rests only on its own conventions. The belief that mathematics is more securely founded if it is reduced to logic is, on the whole, only a misunderstanding. Without being specific, we merely wish to remark that today a new interpretation has become possible. According to it mathematics consists neither of empirical propositions (Mill), nor of synthetic a priori

9. Present Status of the Investigation of the Foundations

judgments (Kant, Hamilton, Poincaré), nor of analytical propositions, respectively, tautologies (Frege, Russell, Ramsey, Hahn, Carnap, etc.), but contains a series of deductive systems, which develop the inferences of arbitrarily chosen assumptions. Logic is only one such calculus which is no more important than the others. This view is essentially the *axiomatic* one, which only has to be supplemented by the disclosure of the dependence that exists between the mathematical symbols and the meanings of words in the colloquial language. The investigation of this dependence belongs to the domain of general logical grammar, which in these days takes great strides toward promising formulations. The question here is to set up exact rules for the application of the numerals and the equality sign within the language. In doing this all the facts will have to be taken into account which we have cited with regard to the application of the word "numerically equivalent" and the numerals.

As to the meaning of an equation we only wish to mention that $a = b$ is used in mathematics as a *rule* which expresses that a, wherever it appears, can be replaced by b. Again, in the context of language $2 + 2 = 4$ is a rule for drawing conclusions, which serves as a transition between propositions. Thus I can conclude from the propositions "I have 2 shillings in my left pocket" and "I have 2 shillings in my right pocket" that I have 4 shillings in both pockets. Therefore the equation does not correspond to a tautology, but to a direction. (These two must be carefully distinguished also in logic.) The equation contains in certain senses a predicative element and stands much closer to an empirical proposition than to a tautology. It is just a rule which guides our behavior (similar to a rule of moving in chess), which we will comply with or transgress. If the equation were a tautology, all this would be impossible. For what should it mean to comply with or to infringe a tautology?

This much is certain. Today we can no longer hold the view that the totality of mathematics is a necessary consequence of a few propositions, such as the five axioms of Peano, and that the truth of the entire system is secured if the truth of the five basic propositions is established. Mathematics is not *one* system but a multitude of systems; we must, so to speak, always begin to construct anew. Consequently, the attempt to reduce Peano's axioms to purely logical propositions loses much of its value.

We will say that the propositions of arithmetic are neither true nor false, but only compatible or non-compatible with certain conventions. By this consideration we overcome a certain dualism which today governs the investigation of the foundations of mathematics. We are referring to the situation wherein the laws of natural numbers are believed to express eternal and irrefutable truths; while the laws of the integers, rational and real numbers are regarded as mere conventions. This is a half-hearted attitude, an inner inconsistency, and the entire foregoing development of arithmetic shows which way we have to go.

The reader will ask: "If the basic laws of arithmetic are arbitrary assumptions, can't they be changed and a new arithmetic attained?" Certainly! The possibility of the "number series 1, 2, 3, 4, 5, many" has already been mentioned. Let us imagine a segment being divided into parts by points. Then it means something to say that the segment has 2, 3, 4, ... parts but not that it has one part. In this case we would much rather count

$$0, 2, 3, 4, \ldots$$

and this corresponds to the series of propositions: "The segment is undivided," "The segment is divided into two parts," etc. This means that here we do not count the parts according to the scheme which we use otherwise. This case gives us an idea of number series in which certain numbers are missing, but whose absence will actually not be noticed. It would be interesting to invent an arithmetic and a possible application of

9. Present Status of the Investigation of the Foundations

which lacked the number 5, without the omission of this number being perceived as a defect.

However, not only can we think of the number series as changed but also the operations. Let us imagine that we are performing an addition of many million numbers. The results of two computers will most probably not agree in such a case. We now ask: Does the concept of probability thereby enter arithmetic? Is computing now a kind of experiment? It seems that two standpoints can here be taken. In one case the rules of addition are held fast and we say that one of the computers must have made a mistake. In the other case, a new calculus is introduced which is only similar to addition, a calculus in which we can no longer speak of a definite sum, or where we must say that the sum lies between this and that limit. The pursuit of such possibilities would be very useful for the philosophy of arithmetic. For we would then seem to see our arithmetic actually set off against a background of other and related calculi, and understand more clearly to which facts of reality our arithmetic is adapted.

Now we seem to have a fairly full view of the error incident to the thesis of the logical school. It was believed that by reducing arithmetic to logic a firm foundation was being built under arithmetic. The laws of logic were indeed held as the safest of all truths, as "cornerstones, fastened in an eternal foundation, certainly inundable by our thoughts but not moveable," (Frege). A purely logical definition of the number concept would thus be made the starting point for rigorous deductions which reduce the truths of arithmetic to the deeper ones of logic. However, the expression "to found arithmetic" gives us a false picture, because it gives us the idea that its structure is to be erected on certain basic truths. Instead, arithmetic is a calculus which starts only from certain conventions but floats as freely as the solar system and rests on nothing.

The conclusion which can be drawn from these considerations is that we can only describe arithmetic, namely, find its

9. Present Status of the Investigation of the Foundations

rules, not give a basis for them. Such a basis could not satisfy us, for the very reason that it must end sometime and then refer to something which can no longer be founded. Only the convention is the ultimate. Anything that looks like a foundation is, strictly speaking, already adulterated and must not satisfy us.

10. Limit and Point of Accumulation

The mathematics of the 17th and 18th centuries bears the impress of calculating with infinite processes. The mathematician of that day, however, had little success in obtaining clear ideas. This may be attested by a problem which was frequently discussed in detail in the 18th century, i.e., the problem of what is the sum of the infinite series

$$1 - 1 + 1 - 1 + 1 - 1 + 1 - 1 + \ldots$$

In 1703 the Camaldolite monk Guido Grandi published a work wherein he used the geometric series

$$1 + x + x^2 + x^3 + x^4 \ldots$$

which has the sum $\frac{1}{1-x}$ to obtain for $x = -1$ the result

$$1 - 1 + 1 - 1 + 1 - 1 + \ldots = \tfrac{1}{2}.$$

In a publication appearing seven years later, which is dedicated to the *Deo veritatis, luminum patri, scientiarum domino, geometriae praesidi*, he came back to this result and sought to justify it by narrating a parable: In a division of a paternal legacy two brothers inherit a stone of inestimable value. As the will forbids them to sell the stone, they agree among themselves to deposit it alternately for a year at a time in each other's museum. If we now stipulate that this understanding shall be valid between the two families for all eternity, then the families

10. LIMIT AND POINT OF ACCUMULATION

of each brother will be given the stone infinitely often, and have it taken away infinitely often, and still each has the half possession of the stone. Incidentally, it is worthy of note that Grandi concluded from the formula

$$1 - 1 + 1 - 1 + 1 - 1 + \ldots = \tfrac{1}{2}$$

by always taking two successive terms at a time that

$$0 + 0 + 0 + 0 + 0 + \ldots = \tfrac{1}{2}$$

wherein he found a proof of the possibility of the creation of the world from nothing.[1]

The publication of Grandi prompted a discussion between Leibniz and Wolff, Grandi and Varignon. Wolff asked Leibniz what he thought of the remarkable things which appear in Grandi's book, and Leibniz offered an opinion in a letter published 1713 in the *Acta eruditorum* (*Journal of Savants*). He does not at all agree with Grandi's juridical explanation; however, regarding the point in question he agrees with him and seeks to support his result by the following argument: if the series is broken off after an even number of terms, the sum is 0; if it is broken off after an odd number of terms, the sum is 1. But the series is infinite. Therefore, since we can ascribe to infinity neither the character of an even nor that of an odd number, the series can have neither the sum 1 nor the sum 0. Now the theory of probability tells us that if two values of a quantity are equally likely their arithmetic mean must be taken as the true value. Hence we must rightfully attribute the value $\dfrac{0+1}{2} = \tfrac{1}{2}$ to the series. Though this kind of argument, adds Leibniz, seems to be more metaphysical than mathematical, it is nevertheless completely safe.

---

[1] These and the following historical remarks are due to Reif*, Geschichte der unendlichen Reihen* (*History of Infinite Series*), 188*

10. Limit and Point of Accumulation

Relying essentially on the authority of Leibniz, Euler later thought that every infinite series must have a definite sum; Goldbach, Daniel Bernoulli and other important mathematicians living in Euler's time also shared this view. What's more, Euler explicitly took (in a letter to Goldbach 1745) the value of an infinite series to be equal to the value of the analytic function whose expansion gives rise to the series. The deceptiveness of this conviction is evidenced by the fact that there are distinct functions whose expansion produces the same series. For example,

$$\frac{1+x}{1+x+x^2} = 1 - x^2 + x^3 - x^5 + x^6 - x^8 + \ldots$$

is a series which again leads to Grandi's series for $x = 1$; we would therefore be able to "prove" that

$$1 - 1 + 1 - 1 + 1 - 1 + \ldots = \tfrac{2}{3}$$

just as well as $\tfrac{1}{2}$. Hence what is the true value of this series?

On analyzing this question it was recognized to be a typical pseudo-problem. One must first *define* what is meant by the sum of an infinite series; for at the outset the word "sum" is explained only for finitely many numbers. In so doing, it turns out that the definition of the sum of an infinite series leads to a more fundamental concept, namely, to that of *limiting value*. We will now consider this concept.

What is meant by a limiting value? We will first illustrate it by an example. For this purpose let us take the expression $\frac{n-1}{n}$ and set successively for n the values $n = 1, 2, 3, 4, \ldots$ We thereby obtain $0, \tfrac{1}{2}, \tfrac{2}{3}, \tfrac{3}{4}, \ldots$, and this sequence of numbers comes closer and closer to the value 1 the further it is carried out. Now it was thought that our expression actually assumed the value 1 for $n = \infty$ and thereby infinity itself was unhesitatingly regarded as a number. This conception cannot stand up before criticism. For what shall we mean, strictly

10. LIMIT AND POINT OF ACCUMULATION

speaking, by an infinite number? It is obvious that infinity is not a clear-cut number entity such as, let us say, the number 5. Attributes such as even, odd, prime number, divisible, etc., cannot be applied to infinity. Therefore it has no precise sense to say that n is given the "value infinity." On the contrary, in our example n indicates a finite number which continually increases; n *is* so to speak never infinite, but only *becomes* infinite. A state of becoming which never comes to an end is always incident to the concept of infinity. This concept of infinity is also called the potential infinite (to distinguish it from the actual infinite, which means an infinite totality).— The formula therefore never yields 1. It would not make matters clearer to say that the terms of the sequence gradually approach 1, for the terms of the sequence *are* actually never 1

Instead, it will be more correct to say that the expression $\frac{n-1}{n}$ tends boundlessly to 1 as n increases without end and that the number 1 will be regarded as the limiting value or the limit of this sequence. The limiting value is therefore conceptually a new entity, something that is added to the stock of numbers actually occurring in the sequence and which only has a definite relationship to them. Since the time of Cauchy, we indicate this relationship by the notation

$$\lim_{n \to \infty} \frac{n-1}{n} = 1.$$

By formulating the concept of limiting value a little differently, we finally obtain the view which prevails today. Instead of saying that the expression $\frac{n-1}{n}$ has the limiting value 1, we can evidently also say that the sequence of numbers 0, ½, ⅔, ¾, ... differs from 1 less and less the further it is continued, in other words, the difference between 1 and $\frac{n-1}{n}$

10. LIMIT AND POINT OF ACCUMULATION

becomes smaller and smaller as the number n increases. In the brief but extremely pertinent symbolic language of mathematics we express this as follows: the difference $1 - \frac{n-1}{n}$ will become eventually less than any value ε, no matter how small; this means

$$1 - \frac{n-1}{n} < \varepsilon,$$

provided that n exceeds a certain value N, i.e., provided that

$$n > N$$

For example, in order that the difference should become less than 0.002 it is sufficient to choose $N = 500$. This means that from the 500th place onward the terms of the sequence differ from the limiting value 1 by less than 0.002. The smaller I choose ε, that is, the closer the terms of the sequence are to draw near to 1, the larger n will become, that is, the further I must stride out in the number sequence.

The relation just described between the numbers ε and N expresses exactly the same thing as the statement $\lim_{n \to \infty} \frac{n-1}{n} = 1$; it actually replaces this statement. On comparison we may at first fail to see the advantages of the second statement over the first; instead we could say that it is much more complicated. However, there is one fact which is very decidedly in its favor: *infinity no longer occurs in it*; on the contrary, it is a system of relations which refer throughout only to finite quantities. From this example there arises an insight of extraordinary import which can be expressed as follows: if the concept "infinity" occurs in a mathematical statement (in the sense of the potential infinite), the same circumstance can also be described by a system of statements which deal only with relations between finite numbers. Consequently the expression "infinity" could be entirely banished from the vocabulary of

10. LIMIT AND POINT OF ACCUMULATION

mathematics without thereby sacrificing the minutest part of the content of its laws. Indeed a special attraction may even be found in constructing an infinitesimal calculus in which the concept of infinity is not even applied once, indirectly or directly. There is no doubt that such a construction is possible; however, hardly any attempt has been made to do this; rather there are reasons to keep carrying infinity along in the formulas, but not in the sense of a primitive concept. Thus in calculations the formula

$$\lim_{n \to \infty} \frac{n-1}{n} = 1$$

proves to be so much more convenient and appropriate than the complicated system of inequalities that it is advantageous to retain infinity as a *facon de parler*. This concept therefore is a symbol which we use in order to describe certain intricate relations in a concise and clear manner.

We will now give a general formulation of the concept of limiting value. The nonending sequence of numbers

$$a_1, a_2, a_3, \ldots a_n, \ldots$$

is said *to converge to the limiting value* a, in symbols

$$\lim_{n \to \infty} a_n = a,$$

if the sequence of numbers, *continued sufficiently far*, draws *arbitrarily* near to a, that is, if

$$a - a_n < \varepsilon,$$

provided that $n > N$. This formulation says that no matter how small a "threshold value" ε is given, I can always find an index N so that the terms

$$a_{N+1}, a_{N+2}, a_{N+3} \ldots$$

deviate from the limiting value a by less than ε. A sequence which does not converge is called divergent.

10. LIMIT AND POINT OF ACCUMULATION

Only one point remains to be considered. As it stands, our criterion still does not express all we wish to say. Thus, instead of the limiting value a let us take any other number a*, where a* is only smaller than all numbers of the sequence. Then the criterion would be satisfied for this number a*, since a* — a_n is a negative number, whatever n may be, and therefore certainly smaller than ε. Hence the criterion merely expresses that a is surpassed by all numbers of the sequence, but not that a is the limiting value of the sequence. However it is easy to remedy this situation for we only have to add that the *absolute value* (which means: without regard to sign) of the difference a — a_n shall become arbitrarily small. This is indicated by writing the criterion in the form

$$|a - a_n| < \varepsilon \text{ for } n > N.$$

N is thereby dependent on ε; the smaller ε is chosen, the larger in general the number N must be taken; if we wish to emphasize this dependence, the second inequality can be written in the form

$$n > N(\varepsilon).$$

This entire discussion is founded on the concept of sequence of numbers, which we will now consider somewhat more exactly. A sequence of numbers is generated by replacing the terms of the sequence 1, 2, 3, 4, ... by any other numbers; 1 by a_1, 2 by a_2, etc. But what does this mean? Are we proposing that this substitution be carried out at will? Shall we allow the possibility of infinitely many choices? Or shall we demand that the terms of the sequence be formed according to a *law*? There is no agreement regarding these points. Many mathematicians will expressly also allow sequences without any law. This view leads to certain difficulties which we will set forth later. Actually, at any rate, only sequences are used for which a definite law of formation is known. By such a law we under-

10. LIMIT AND POINT OF ACCUMULATION

stand either a formula, for example $a_n = \frac{n-1}{n}$, or a description in words: the law of formation of the formula

$$2, 3, 5, 7, 11, 13, 17, \ldots$$

can only be expressed by saying that the n-th prime shall be at the n-th position. Today no formula is known by which the prime numbers can be computed. Hence the verbally stated law cannot be converted into a formula. Nevertheless, a clear rule is submitted for the formation of a sequence of numbers. The essential point is that this rule assigns to every natural number n a very definite number a_n.

There is a further difficulty. According to our explanation a sequence is determined as soon as we know a *law* by which the terms of the sequence are formed. This explanation would then be precise and clear if the same could be said of the concept of law. But can it really be said? On considering the facts somewhat more carefully, we are immediately beset with doubts. Let us form a sequence of numbers $t_1, t_2, t_3, \ldots$ by the following rules: the number t_n shall be 1 if three integers x, y, z can be found satisfying

$$x^n + y^n = z^n$$

while $t_n = 0$ if no integers can be found which satisfy this equation. The sequence therefore begins as follows:

$$1, 1, 0, 0, 0, \ldots$$

and today nobody can say whether or not the first two terms are followed only by zeros. This decision depends on the solution of a famous number-theoretical problem. If Fermat, after whom it is named, made the correct conjecture, only zeros follow; if he is wrong, then ones will appear somewhere. Today it is only proved that Fermat's statement is correct for all numbers n in the first hundred and for many larger numbers. Hence our sequence can be continued for numbers up to 100 from then on, however, its appearance becomes uncertain. Since it is not at all certain that this problem has a solution, it is possible that we will never know how the sequence runs on

10. LIMIT AND POINT OF ACCUMULATION

Now is our rule a law? Or will it stand for a law only after Fermat's problem has been solved?

Consequently the concept of sequence of numbers is not free of difficulties. However, for the present we will let this point pass and try to orient ourselves a little in the domain of sequences. We will first consider some sequences of numbers and make a guess at the law by which they are formed.

1. $1, 2, 3, 4, \ldots$ $a_n = n$
2. $2, 4, 8, 16, \ldots$ $a_n = 2^n$
3. $-1, 2, -3, 4, \ldots$ $a_n = (-1)^n \cdot n$
4. $3, 3, 3, 3, \ldots$ $a_n = 3$
5. $1, \frac{1}{2}, \frac{1}{3}, \frac{1}{4}, \ldots$ $a_n = \frac{1}{n}$
6. $0, \frac{1}{2}, \frac{2}{3}, \frac{3}{4}, \ldots$ $a_n = \frac{n-1}{n}$
7. $0, \frac{2}{2}, \frac{2}{3}, \frac{4}{4}, \frac{4}{5}, \ldots$ $a_n = \frac{n + (-1)^n}{n}$
8. $\frac{1}{1}, \frac{1}{2}, \frac{2}{3}, \frac{3}{5}, \frac{5}{8}, \frac{8}{13}, \ldots$ the law of this sequence can only be expressed by a recursive formula; if $\frac{p_n}{q_n}$ designates the n-th term of the sequence, then $p_1 = 1$, $q_1 = 1$ and $p_{n+1} = q_n$

$$q_{n+1} = p_n + q_n.$$

9. $3, \frac{1}{2}, 2\frac{1}{3}, \frac{1}{4}, 2\frac{1}{5}, \frac{1}{6} \ldots a_n = \frac{1}{n}[n - (-1)^n \cdot n + 1].$
10. $0, 1, 0, 1, 0, 1, \ldots$ $a_n = \frac{1}{2}[1 + (-1)^n].$

These ten sequences show great differences of structure. A clear picture is gained if we represent them geometrically. Then the accompanying graphs result (Fig. 11). The most important difference which meets the eye is this: some of the sequences are dispersed to infinity (1 to 3); others (5 to 8) cluster about a definite point, converge to this point (in example 5, it is the point 0; in examples 6 and 7, the point 1; in example 8 a point which lies at any rate between $\frac{1}{2}$ and $\frac{2}{3}$ and that is all we know about it); in some cases, the approach to the limit point takes place through numbers which increase

10. LIMIT AND POINT OF ACCUMULATION

throughout, i.e., from the left as in (6); in some cases, through numbers which decrease throughout, i.e., from the right as in (5); and in some cases, from both sides (7 and 8): i.e., the terms of the sequence hop, as it were, about the limiting value and enclose it in ever narrower intervals. Finally, in a third kind of sequence we can recognize more than one center of condensation, i.e., points about which the sequence clusters. For example, the sequence 9 condenses about the points 0 and 2. The remaining examples consist in the nonending repetition of one or more points. We will return to this later on.

In studying sequences it is a very important matter whether there are points on the number axis where they seem to condense, that is, to be infinitely densely clustered. We will call such points *points of accumulation* and will try to set up a rigorous definition for the intuitable behavior just described.

When will we call a point a point of accumulation? Evidently, if in the immediate neighborhood (to the right or to the left or on both sides) of such a point there is a tremendous crowd of terms of the sequence. Hence we define a point of accumulation of the sequence to be a point such that every interval

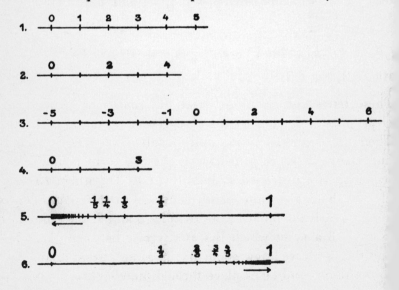

10. Limit and Point of Accumulation

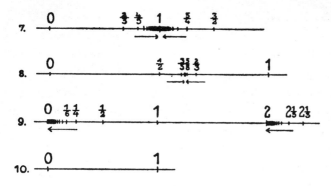

Figure 11

no matter how small, constructed around this point, already contains infinitely many terms of the sequence.

It will be very profitable for the reader to use intuitive devices to familiarize himself with the sense of this definition. For this purpose let us visualize the points of the sequence, let us say, as beads strung on a thread. Then the definition says the following: if a piece of thread, no matter how small, is removed which contains a point of accumulation in its interior, then an infinite number of beads has also been removed at the same time.

Consequently, a limiting point is always a point of accumulation; however, the converse is not true, namely, a point of accumulation need not be a limiting value (see example 9). Now, how is the exact relation between these concepts to be formulated? We will see how this is to be done if we look for the properties that these two concepts have in common and those which they do not. In both cases any interval, no matter how small, that includes such a point, contains infinitely many points of the sequence. The difference shows up as soon as the terms which remain outside of the interval are taken into consideration: in the case of a limiting value there are only *finitely many*, in the case of a point of accumulation (which is not at the same time a limiting value) there are *infinitely many*. For instance, in example 9 if the point of accumulation 2 is

10. LIMIT AND POINT OF ACCUMULATION

included in a small interval, say with the radius 1/1000 (that is, the interval from 1.999 to 2.001), then this interval certainly contains infinitely many points of the sequence; but outside of this interval there also lie infinitely many others, namely, all those points which are clustered around the point of accumulation 0. On the contrary, if the sequence has a limit, then every interval which contains the limit also includes "nearly all" terms of the sequence (namely, all with the exception of at most finitely many). In terms of the expression "nearly all"[2] the wording of the two definitions can be put into a rather similar form.

A *point of accumulation* is a point such that every neighborhood of this point, no matter how small, contains *infinitely many* terms of the sequence.

A *limit* is a point such that every neighborhood of this point, no matter how small, contains *nearly all* terms of the sequence.

However, if the sequence (as in examples 4 and 10) consists of the nonending repetition of one or more terms, these definitions no longer hold. Since it is useful for the calculus to be able to say that a sequence has a limiting value or a point of accumulation even in such cases we extend these concepts a little further. Accordingly we define:

a is called a point of accumulation of the sequence $a_1, a_2, a_3, \ldots,$ if for *infinitely many* n

$$|a - a_n| < \varepsilon;$$

a is called a limiting value of the sequence $a_1, a_2, a_3 \ldots$ if for *nearly all* n

$$|a - a_n| < \varepsilon,$$

where ε denotes an arbitrarily small positive number.

This cursory examination has already unfolded extraordinary differences: sequences which have no points of accumulation

[2] It is due to G. Kowalewski.

10. LIMIT AND POINT OF ACCUMULATION

tion, those which exhibit exactly one point of accumulation, and those which have many points of accumulation.

Perhaps the reader thinks that sequences with exactly one point of accumulation are the same as convergent sequences. This would be a mistake. Let us consider the sequence

$$1, \tfrac{1}{2}, 3, \tfrac{1}{4}, 5, \tfrac{1}{6}, 7, \tfrac{1}{8}, \ldots$$

for every odd n, $a_n = n$; for every even n, $a_n = 1/n$. Obviously this sequence has the single point of accumulation 0; nevertheless it is not a limiting value, since this sequence contains infinitely many other terms which are scattered unboundedly. The fact that this sequence consists, properly speaking, of two distinct sequences which are pieced together in a purely external manner should not be a matter of argument. For, in the first place we have stated a clear law for the formation of this sequence; and secondly we can even express the law of formation by a single formula, namely, by

$$a_n = \left(\frac{1}{n}\right)^{(-1)^n}$$

This example illustrates that a point of accumulation, even if there is only one such point, does not have to be a limit. If a sequence is to possess a limit, it must satisfy more requirements than if it is only to have a point of accumulation.

We will now give an example of a sequence which has infinitely many points of accumulation. For this purpose we consider the totality of numbers of the form $\frac{1}{m} + \frac{1}{n}$ and arrange them as shown in the following table:

10. LIMIT AND POINT OF ACCUMULATION

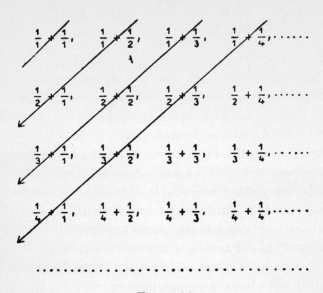

Figure 12

The first row is generated by setting m = 1 and letting n run through the sequence of natural numbers; the second row by setting m = 2 and letting n behave as above; and so on. It is clear that our scheme will contain every fraction of the form $\frac{1}{m} + \frac{1}{n}$; for example the fraction $\frac{1}{25} + \frac{1}{39}$ is in the 25th row and 39th column. We will now fuse these infinitely many sequences into a single one by running through the terms of the table as indicated by the arrows. We then obtain a sequence which exhibits every fraction of the required form. Actually each fraction occurs twice, namely, as $\frac{1}{m} + \frac{1}{n}$ and as $\frac{1}{n} + \frac{1}{m}$; if we wish to avoid repetitions, then the fractions which have already occurred once can be deleted. The sequence thereby obtained has *infinitely many points of accumulation*. Thus the numbers of the first row converge to 1, the numbers of the second row to ½, the numbers of the third to ⅓, etc., and 1, ½, ⅓, . . . will consequently be points of accumulation of the whole sequence. The following graph may represent the appearance of this sequence:

10. Limit and Point of Accumulation

Figure 13

The graph shows how the points of accumulation cluster more and more densely toward 0: 0 is itself a point of accumulation of the points of accumulation.

Sequences can even be constructed with the paradoxical property that *each* point is a point of accumulation. The totality of rational numbers offers an example of such a sequence. In order to arrange them in a sequence we first think of all fractions with numerator 1 as written down, then all fractions with the numerator 2, after this all with the numerator 3, etc. The positive rational numbers can then be written down by means of the following two dimensional scheme:

$$\frac{1}{1}, \quad \frac{1}{2}, \quad \frac{1}{3}, \quad \frac{1}{4}, \ldots$$

$$\frac{2}{1}, \quad \frac{2}{2}, \quad \frac{2}{3}, \quad \frac{2}{4}, \ldots$$

$$\frac{3}{1}, \quad \frac{3}{2}, \quad \frac{3}{3}, \quad \frac{3}{4}, \ldots$$

$$\frac{4}{1}, \quad \frac{4}{2}, \quad \frac{4}{3}, \quad \frac{4}{4}, \ldots$$

. . . .

Figure 14

If the fractions are arranged, let us say, by filing the diagonals one after another (as in Figure 12), a sequence is obtained which begins with:

$$\tfrac{1}{1}, \tfrac{1}{2}, \tfrac{2}{1}, \tfrac{1}{3}, \tfrac{2}{2}, \tfrac{3}{1}, \tfrac{1}{4}, \tfrac{2}{3}, \tfrac{3}{2}, \tfrac{4}{1}, \ldots$$

Of course in this sequence each fraction will occur many times, for example, 1/1, 2/2. If we again omit every fraction which has already occurred once, an ordering is obtained which ex-

10. LIMIT AND POINT OF ACCUMULATION

hibits each fraction only once and which pushes on eventually to any fraction. In this way the set of positive rational numbers is arranged in a sequence; true enough, this was done at the price of thoroughly destroying the natural ordering of the fractions according to magnitude. Now to include all fractions, we only have to insert each negative fraction after the corresponding fraction of the above sequence and to set 0 before the first term. In this sequence *every* rational number is a point of accumulation, so that the sequence consists only of points of accumulation. Furthermore, the irrational points are also points of accumulation (which contain in every neighborhood infinitely many rational numbers), so that the sequence has more accumulation points than terms.—

Now that we have obtained a first orientation in this subject matter we can return to the question of what is the sum of the series

$$1 - 1 + 1 - 1 + 1 - 1 + \ldots$$

We have already mentioned that in arithmetic the concept of sum is defined only for finitely many numbers. If we wish to speak of the sum of an infinite series, the expression "sum" must be defined anew. How shall we do this? The following definition seems most natural. Instead of adding all terms of a given infinite series

$$a_1 + a_2 + a_3 + \ldots + a_n + \ldots$$

which we cannot do, we will first form the sum of only the first two terms, then of the first three terms, etc. The partial sums thereby obtained

$$s_2 = a_1 + a_2$$
$$s_3 = a_1 + a_2 + a_3$$
$$\text{etc.}$$

form a sequence $s_2, s_3, \ldots s_n, \ldots$, and we now look to see whether this sequence tends to a definite limiting value. If it does, then we will call the limiting value the sum of the infinite series. In other words, the sum s of the infinite series

$$a_1 + a_2 + a_3 + \ldots + a_n + \ldots$$

is defined by the expression

$$s = \lim_{n \to \infty} s_n.$$

It is well to note that s is not a sum in the sense in which the word was previously defined, but the *limiting value of a sequence of sums*. Hence it is not to be expected as a matter of course that all properties of finite sums will be indiscriminately valid also for the infinite structures. A finite series of numbers always has a sum; for infinite series this is by no means valid, since the sequence of the s_n need not converge. The series $1 - 1 + 1 - 1 + \ldots$ is in fact an example of this. Namely, if the sequence of the partial sums is formed, then these are alternately 0 and 1; the sequence of the numbers

$$0, 1, 0, 1, 0, 1, \ldots$$

does not tend to a fixed limit; and we must therefore say that our series does not have a sum.

However we must never forget that this result depends on our definition of sum and therefore on something arbitrary. An infinite series does not have a sum by nature; rather it acquires a sum only if we associate one to it by a procedure. Now the concept of sum could also be set up in a different way and then it could happen that our series would turn out to be summable. For example, such a definition could be the following: since the partial sums of our series oscillate back and forth between 0 and 1, the idea suggests itself to form their arithmetic means

$$S_1 = \frac{s_1}{1}, \ S_2 = \frac{s_1 + s_2}{2}, \ S_3 = \frac{s_1 + s_2 + s_3}{3}, \ \ldots$$

in the expectation that these tend to a limiting value. If this the case, then we will call this limiting value the sum of the infinite series. Actually this process yields the sequence

10. LIMIT AND POINT OF ACCUMULATION

$$\frac{1}{1}, \frac{1}{2}, \frac{2}{3}, \frac{2}{4}, \frac{3}{5}, \frac{3}{6}, \frac{4}{7}, \frac{4}{8}, \ldots,$$

which obviously approaches the limiting value ½. This new definition of sum imparts a clear meaning to the statement of Grandi and Leibniz; to be sure a meaning which was still unknown to those men.

If we accept this definition, then the following can be proved: every series convergent in the sense defined previously is also convergent in the new sense and yields the same sum. On the other hand, series which are divergent in the old sense could turn out to be convergent (summable) in the new sense, so that the new process has a greater ability to form sums than the old. Today a great number of processes for summing divergent series are known, processes which can partly be ordered in a scale such that every successive process has a larger operational field, i.e., allows more series to be summed than the preceding one. The formation of these processes is again ruled by the requirements of permanence: a series summable in the sense of the process V_k shall also be summable in the sense of the process V_{k+1} and yield the same sum. Such considerations show very impressively that it actually depends only on the definition whether or not an infinite series has a sum

11. Operating with Sequences. Differential Quotient

In this section we will be concerned with the possibility of operating with infinite sequences. Addition may be used to show that this is feasible. If two sequences

$$a_1, a_2, a_3, \ldots\ldots a_n, \ldots$$
$$b_1, b_2, b_3, \ldots\ldots b_n, \ldots$$

are given which converge to the limiting values a, or b, respectively, it is natural to think of their sum as the sequence

$$a_1 + b_1, a_2 + b_2, \ldots\ldots a_n + b_n, \ldots$$

Actually, this sequence is again convergent and has a + b as its limiting value. Before proving this rigorously, we wish to make the statement plausible by an example. Let us assume that the first sequence converges to 1, the second to 2. This means that after a while the terms of the first sequence differ from 1 only to an inperceptible extent, the terms of the second imperceptibly from 2. But then the sum of two such terms will deviate only a little from 3. This is precisely the content of the statement.

The rigorous proof uses the relation

$$|a+b| \leq |a| + |b|;$$

namely, if a and b have the same sign, then

$$|a+b| = |a| + |b|;$$

but if, for example, a is positive and b is negative, then

$$|a+b| < |a| + |b|.$$

11. Operating with Sequences. Differential Quotient

In general, therefore, no more can be stated about the absolute value of a sum than that it is *smaller* or *at most equal* to the sum of the absolute values.

To obtain the proof itself we only have to set up rigorously the intuitive line of reasoning used above. Let us assume that the sequences (a_n) and (b_n) converge to a, or b, respectively, that is,

$$|a - a_n| < \varepsilon, \text{ for } n > N_1$$
$$|b - b_n| < \varepsilon, \text{ for } n > N_2.$$

The numbers N_1 and N_2 will in general be different, since they depend entirely on the rate at which the sequences tend toward their limits. (If the first sequence converges rapidly, just a few terms will suffice in order to attain a value less than ε units away from the limit; if the second converges slowly, then many more terms are necessary to reach the same value.) If N denotes the larger of the two numbers N_1 and N_2, then we have in each case

$$|a - a_n| < \varepsilon, \text{ provided that } n > N$$
$$|b - b_n| < \varepsilon, \text{ provided that } n > N$$

Hence from the N-th term on, each of the two sequences will deviate from their limiting value by less than ε. Now if the sequence $(a_n + b_n)$ is to converge to $a + b$, the expression

$$|(a + b) - (a_n + b_n)|$$

must become arbitrarily small as n increases. But

$$|(a + b) - (a_n + b_n)| = |(a - a_n) + (b - b_n)|$$

and this expression, as a consequence of our preliminary remark, is

$$\leq |a - a_n| + |b - b_n|$$
$$< \varepsilon + \varepsilon = 2\varepsilon,$$

and therefore actually becomes arbitrarily small with increasing n.

11. OPERATING WITH SEQUENCES. DIFFERENTIAL QUOTIENT

The theorem just proved can also be stated as follows: the limiting value of a sum of two sequences is equal to the sum of the limiting values of the sequences; in symbols,

(1) $$\lim (a_n + b_n) = \lim a_n + \lim b_n.$$

In an analogous manner three further formulae can now be established.

(2) $\qquad \lim (a_n - b_n) = \lim a_n - \lim b_n$
(3) $\qquad \lim (a_n \cdot b_n) = \lim a_n \cdot \lim b_n$
(4) $\qquad \lim (a_n : b_n) = \lim a_n : \lim b_n.$

The last formula is valid only if $\lim b_n$ is different from 0 (and this corresponds exactly to the proposition that one must not divide by 0). Hence *addition, subtraction, multiplication, division do not lead us out of the domain of convergent sequences.* In other words, the convergent sequence form a *field*.

Our four formulae can be regarded from a somewhat more general point of view if we decide to look upon the passage to the limit as a new operation, as *limit operation*, which now takes its place besides the four basic operations of arithmetic. Our four formulae can then be comprised in the one proposition: *the limit operation is commutative with the four basic operations of arithmetic.* At first it may appear strange that we speak here of an operation. The following discussion, however, will make it plausible. Just as addition is a process which leads from a series of given numbers, the summands, to a new number, the sum, so also does the passage to the limit stand for a process which leads from the infinitely many numbers of a sequence to a new number, the limit. The word "operation" expresses this analogy. The question whether the passage to the limit is "in reality" an operation is idle, since we have not sharply defined the concept "operation"; the word is disposed of just by analogy.

The analogy with the operations of arithmetic can still be pushed a little further. As we know, diverse facts can be discovered about a division—for example, whether it has the

143

11. OPERATING WITH SEQUENCES. DIFFERENTIAL QUOTIENT

same result as another—without actually carrying it out. And therefore by examining two sequences, we can also tell whether they tend to the same limit without having to know this limit. The execution of these ideas presupposes, however, some concepts which we now will take up. A sequence which converges to 0 is called a "null sequence." The criterion for a null sequence is

$$|a_n| < \varepsilon, \text{ if only } n > N,$$

that is, if the terms decrease unboundedly. A convergent sequence which is not a null sequence will have either a positive or a negative limiting value. In the first case their terms cluster about a point on the axis of positive numbers. This means that if a point c is marked somewhere between this limit and 0, almost all terms of the sequence will lie to the right of c. Consequently, a series is positive if a positive number c can be exhibited which will be surpassed by almost all terms of the sequence. The sequence is negative if a negative number — c exists which will surpass almost all terms of the sequence.

In terms of these concepts we can next formulate the following three propositions:

1. The sequence (a_n) yields a larger limit than the sequence (b_n) if the difference sequence $(a_n - b_n)$ is positive.

2. The sequence (a_n) yields a smaller limit than the sequence (b_n) if the difference sequence $(a_n - b_n)$ is negative.

3. The sequence (a_n) yields the same limit as the sequence (b_n) if the difference sequence $(a_n - b_n)$ is a null sequence.

It may be sufficient to prove only the first proposition. If the difference sequence $(a_n - b_n)$ is positive, then lim $(a_n - b_n) > 0$; but by formula (2) this is the same as saying that $\lim a_n - \lim b_n > 0$, that is, $\lim a_n > \lim b_n$.

An advantage of this formulation lies in the fact that it is often much easier to discover something about the behavior of difference sequences than about the limiting values them

11. OPERATING WITH SEQUENCES. DIFFERENTIAL QUOTIENT

selves. However, our main reason for quoting these propositions is that later on they are to serve as a model for the rigorous construction of the real numbers. This means that we will construct a calculus with sequences and will copy the definitions for "greater," "equal," "smaller," "sum," "difference," etc., after the propositions proved above.

We have said that we are not allowed to divide by a null sequence, because a sequence can thereby arise which no longer converges. However, nothing prevents us from extending the concept of convergence, so that such sequences are also called convergent. Indeed, such an extension actually forces itself on us as soon as we think of the representation of numbers on the circumference of a circle.[1] If we map, for instance, the sequence

$$0, 1, 2, 3, 4, \ldots$$

on the circle, then the image points crowd closer and closer to N. Any neighborhood, no matter how small, described around N, contains all terms of the sequence. The point N is therefore the limiting value of the sequence, and we have to say—by transferring this manner of speaking to the number axis—that the original sequence tends to the "improper" limiting value ∞. Thus on accepting the number ∞ we must include the unboundedly increasing sequences among the convergent ones. Then in the case of sequences which are not null sequences we can allow division by a null sequence. For example, if we wish to divide the sequence

$$0, 1/2, 2/3, 3/4, 4/5, \ldots 1 - 1/n \ldots$$

by the null sequence

$$1, 1/2, 1/3, 1/4, 1/5, \ldots 1/n \ldots$$

then the sequence

$$0, 1, 2, 3, 4, 5, \ldots n, \ldots,$$

[1] Cf. Fig. 7, p. 45.

11. OPERATING WITH SEQUENCES. DIFFERENTIAL QUOTIENT

is generated which converges, in the new manner of speaking, to ∞.

We must now devote somewhat more care to the question of what happens if we form the quotient of two null sequences. Here various cases are possible.

1. The sequence of quotients has a finite limiting value. If the two sequences

$$1, 1/3, 1/5, 1/7, \ldots 1/{2n-1}, \ldots$$
$$1/2, 1/4, 1/6, 1/8, \ldots 1/{2n}, \ldots$$

are divided, the first by the second, the sequence

$$2, 4/3, 6/5, 8/7, \ldots 2n/{2n-1}, \ldots$$

is obtained, which obviously approaches the limiting value 1

2. If the sequences are

$$1, 1/2, 1/3, 1/4, \ldots 1/n, \ldots$$
$$1, 1/4, 1/9, 1/{16}, \ldots 1/{n^2}, \ldots,$$

then on dividing the second sequence by the first we obtain

$$1, 1/2, 1/3, 1/4, \ldots 1/n, \ldots$$

and on dividing the first by the second,

$$1, 2, 3, 4, \ldots n, \ldots$$

This shows that the quotient of two null sequences can also be 0 or ∞. In older texts this situation would probably be formulated as follows: the two sequences become infinitely small, but the infinite smallness of the second sequence is of higher order than that of the first. Naturally this only means that the terms of the second sequence tend to 0 at a much faster rate than those of the first, that is, they become arbitrarily small relative to those of the first sequence.

3. There can arise a case which is basically different from all those considered up to now, namely, the sequence obtained by division may not have a limiting value at all, instead it may have only points of accumulation. This can be illustrated by the sequences

11. OPERATING WITH SEQUENCES. DIFFERENTIAL QUOTIENT

$$1, 1/3, 1/9, 1/5, 1/25, 1/7, 1/49, 1/9, 1/81, \ldots$$
$$1/2, 1/4, 1/6, 1/8, 1/10, 1/12, 1/14, 1/16, 1/18, \ldots$$

If the second sequence is divided by the first, we obtain

that is, $\quad 1/2, 3/4, 9/6, 5/8, 25/10, 7/12, 49/14, 9/16, 81/18, \ldots,$
$\quad\quad\quad 1/2, 3/4, 1\,1/2, 5/8, 2\,1/2, 7/12, 3\,1/2, 9/16, 4\,1/2, \ldots.$

This series exhibits the following law: the odd terms always increase by 1, so that they form a sequence whose terms increase beyond all bounds; the even terms, however, form a decreasing sequence whose terms approach closer and closer to $1/2$. The sequence has therefore the two points of accumulation 0 and ∞.

Consequently, as a rule, nothing definite can be stated about the quotient of two null sequences. It can be 0 or ∞ or any other number. It need not exist at all if the sequence generated by division diverges. A special investigation is required to determine which of these cases arises.

The significance of these statements about null sequences is tremendous, since the explanation of a concept which was submerged in obscurity in mathematics for nearly two hundred years rests on them. I am referring to the concept of differential quotient. The problem which gave rise to the differential calculus is the problem of tangents. The reader knows no doubt how to draw a tangent to a circle. If he tries to apply this construction to an ellipse, he will note that it fails. The Greek geometers had invented an appropriate process for this case; however, this also breaks down in the case of the parabola, the curve next in kin. Furthermore, the construction for the tangents of a parabola could again not be transferred to the hyperbola. To be brief, this means that a special construction must be devised for each curve, applicable to this curve alone and to no other. Though perhaps this was inconvenient, it could be tolerated as long as only a few curves were known. The situation, however, changed abruptly with the discovery of

11. OPERATING WITH SEQUENCES. DIFFERENTIAL QUOTIENT

Descartes, i.e., coordinate geometry, for an inestimable number of curves thereby entered the field of vision of the mathematician. There arose the problem to develop a *universal* method by which the behavior of tangents could be studied for any curve whatever. This problem, which Descartes had bequeathed to posterity, was solved a generation later by Newton and Leibniz and they did this at about the same time. Today, the solution of this problem is called differential calculus.

We will next introduce the idea of Leibniz by means of an example. The tangent at the point P is obtained in a roundabout way. First, a secant is drawn through P to any other point P_1 which lies on the curve. Next, as P is held fixed P_1 is allowed to move along the curve near and nearer to P. Then the secant turns slowly around P and tends to a fixed limiting position. This limiting position is the tangent.

Simple and self-evident as this idea appears, it marked a certain revolutionary step. Up to the time of Leibniz there existed in mathematics a deep chasm between secant and tangent. For example, propositions valid for the secants of a circle are totally different from those valid for the tangents of a circle, and it would have occurred to no geometer to establish propositions common to these two kinds of lines. It seems that Leibniz, through his philosophical ideas regarding a continuous connection of all things being, his *loi de continuité*, was led to the view that tangents could be adopted among the secants as limiting cases. At any rate, this kind of vision indicated the hour of birth of differential calculus.

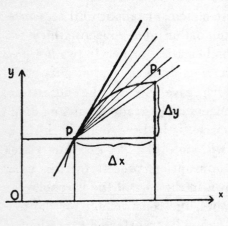

Figure 15

148

11. Operating with Sequences. Differential Quotient

If the passage to the limit is to be carried out computationally, the underlying law of the curve must first be expressed by an equation between the coordinates

$$y = f(x).$$

Let the point P have the coordinates (x, y). If I pass from P to P_1, I must increase the abscissa x by a definite amount, say by Δx; then the ordinate also experiences an increment Δy, which is obviously the surplus of the new ordinate over the old and therefore is equal to $f(x + \Delta x) - f(x)$. The ratio

$$\frac{\Delta y}{\Delta x} = \frac{f(x + \Delta x) - f(x)}{\Delta x}$$

now gives us a measure for the average slope of the curve between the points P and P_1. If we think of the sketch as, let us say, the (exaggerated) view of a mountain road, then the slope in the neighborhood of P is large and decreases gradually from there. On joining the starting and final points by a straight line, the average slope of the route will be represented.

What we seek, however, is not the average slope, but the *instantaneous* slope of the curve at P. For this purpose we will allow the interval which affects our computations to become narrower and narrower, that is, we will continually diminish Δx so as to force the point P_1 to come closer and closer to the point P. If we wish to express this precisely, we will say: if Δx runs through a null sequence, then due to the continuity of the curve Δy also runs through a null sequence. Let us now consider what happens to the quotient $\frac{\Delta x}{\Delta y}$ in the meantime. We believe we see immediately by inspection that this quotient *must* tend to a limiting value, and this limiting value will be precisely the measure of the required instantaneous slope of the curve at the point P. In order to express this we write

$$\lim_{\Delta x \to 0} \frac{\Delta y}{\Delta x} = \frac{dy}{dx}.$$

11. Operating with Sequences. Differential Quotient

$\frac{dy}{dx}$ is called the differential quotient, $\frac{\Delta y}{\Delta x}$ the difference quotient.

However, we must here, from the beginning, dispel an idea which vaguely occurred to the founders of differential calculus and cast a shadow over its creation, namely, the idea that the differential quotient might be a quotient of two infinitely small quantities. On the other hand, it must be clearly understood that the differential quotient is not a quotient at all, but the limiting value of a sequence of quotients. However, this was not yet clear to the founders of differential calculus, although they occasionally came very close to the truth. By and large, they cultivated the view that the differential quotient is the ratio of the quantities $\Delta x, \Delta y$ at the instant at which they just vanish—the *ultima ratio evanescentium incrementorum*, as Newton said. Leibniz and Newton had a feeling that there existed a difficulty in the formation of this concept; however, they were unable to get a really clear idea about it. And their successors, not being critical, and more eagerly after the conquest of new domains than after the clarification of concepts—*Allez en avant, et la foi vous viendra!* (Onward march, and faith will come to you!) d'Alembert is reported to have said—believed, more than ever, that they were calculating with infinitely small quantities, which, lying between 0 and the numbers proper, would form a mysterious intermediate realm.

The reader will ask with some surprise: How does it happen that such important mathematicians did not see through this state of affairs long ago? The confusing feature is that the differential quotient, though not a quotient, behaves in many instances as if it were a quotient. Thus, if y depends on x, then x also depends on y. The differential calculus tells us that the differential quotient of the latter function $\frac{dx}{dy}$ is the reciprocal of $\frac{dy}{dx}$, that is, $\frac{dy}{dx} \cdot \frac{dx}{dy} = 1$, just as if it were a matter

11. OPERATING WITH SEQUENCES. DIFFERENTIAL QUOTIENT

of ordinary fractions. Furthermore, if y is a function of u, and u a function of x, then y through the intermediary u is also a function of x. We now have the law:

$$\frac{dy}{dx} = \frac{dy}{du} \cdot \frac{du}{dx}.$$

After examining this formula closely, we are tempted to say: This is self-evident, the du cancels out anyway. (However, this is an illusion. dy and dx have no meaning at all if taken separately, but only when used in the combination $\frac{dy}{dx}$. This symbol has a meaning only as a whole; in other words, the laws of differential calculus deal with an indecomposable symbol which is merely written in the form of a quotient.) Briefly, it is as if the differential quotient had undertaken to play a prank on the mathematician by behaving just as if it were an actual quotient.

Another difficulty is contained in the concept of differential quotients which we must take into consideration. According to our representation, the differential quotient is the limit to which $\frac{\Delta y}{\Delta x}$ tends as Δx runs through a null sequence. However, we ask, *which* null sequence? There are indeed infinitely many sequences which converge to zero. It is now required that $\frac{\Delta y}{\Delta x}$ converge for *any arbitrary* null sequence and that all limiting values so obtained agree. In other words, we require that the value of the differential quotient remain fixed, independently of the particular manner by which Δx approaches zero. From this standpoint it is no longer self-evident that a differential quotient exists with these properties. In fact, how will one check at all whether the totality of ways of passing to the limit yields the same result? Fortunately the last question is easier to answer than it seems, for in the case of most functions which are dealt with in practice it follows from the very course of the calculation that the result is independent of the particu-

11. OPERATING WITH SEQUENCES. DIFFERENTIAL QUOTIENT

lar choice of null sequence. Sometimes, however, the differential quotient can actually depend on the manner in which the limit is approached and we will next turn our attention to these cases.

12. Remarkable Curves

We now enter a realm which is just as interesting to the epistemologist as to the psychologist: we will study curves which behave so paradoxically that they escape ordinary intuition.

We begin with a case which gives us no trouble at all. Here we have in mind a curve which runs along uninterruptedly and has a corner somewhere. To the question what direction does the curve have at the corner point, we must answer: it has two directions according to whether one moves from the corner to the left or to the right. If we wish to describe this relation by computations, we are led to two distinct formulations according to whether Δx tends to zero through positive or through negative values. In this case we speak of a "right-hand," or a "left-hand differential quotient," respectively.

Next, we will consider a case which is a little more interesting. The reader has certainly already heard something about wave lines, such as those, for instance, an oscillating tuning fork draws upon a sooted drum. The mathematician is accustomed to speak of them as sine curves, since the curve $y = \sin x$ is the representative of this class of curves.

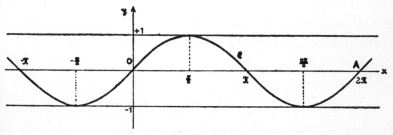

Figure 16

12. REMARKABLE CURVES

x stands for an angle; it is measured not in degrees, minutes and seconds but in radians. This means that if a circle of radius 1 is described about the vertex of the angle the measure of the size of the angle is taken as the length of the arc which the angle cuts out on the circumference of this circle. Thus an angle of 360° is represented by 2π, 180° by π, 90° by $\frac{\pi}{2}$, etc. The reader can see from the figure that the curve $y = \sin x$ contains the wave train OA and that the entire curve is generated by connecting together infinitely many of these wave trains. The sine function assumes its greatest value $+1$ infinitely often, namely, at the points $x = \frac{\pi}{2}, \frac{\pi}{2} \pm 2\pi, \frac{\pi}{2} \pm 4\pi$, etc., and has the value -1 at all points of the form $x = \frac{3\pi}{2} \pm 2n\pi$. The sine curve cuts the x-axis at the points $0, \pm \pi, \pm 2\pi, \pm 3\pi, \ldots\ldots$

The curve we are interested in, however, is not the sine curve but the curve $y = \sin \frac{1}{x}$. Since the sine function takes on all values between $+1$ and -1, this curve must run entirely within the plane strip which extends a distance 1 on both sides of the x-axis. The function takes on its greatest value 1 if $\frac{1}{x}$ is either $\frac{\pi}{2}$ or $\frac{5\pi}{2}$ or $\frac{9\pi}{2}$ ..., which means, if $x = \frac{2}{\pi}, \frac{2}{5\pi}, \frac{2}{9\pi}, \ldots$. At all these points the curve will assume the form of a wave crest, and these points crowd closer and closer about 0. A wave trough is represented at the points $x = \frac{2}{3\pi}, \frac{2}{7\pi}, \frac{2}{11\pi}, \ldots$. For larger values of x, $\frac{1}{x}$ is very small and $\sin \frac{1}{x}$ is almost equal to 0; in other words, the curve draws in closer and closer to the axis of abscissas at points which are way out. From this we conclude that the curve has the following appearance:[1]

---

[1] The curve is drawn out in the figure.

12. REMARKABLE CURVES

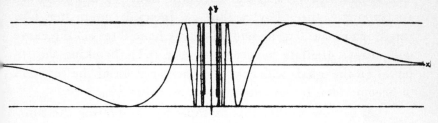

Figure 17

The curve consists of infinitely many wave trains which crowd closer and closer about the origin and visibly climb more steeply. At the point x = 0 itself the function is not defined; any value whatsoever can be ascribed to it at this point, for example, 0 or ½ or any other number. Can the course of this curve be clearly visualized? What can be visualized intuitively are the individual wave trains and perhaps also the way in which they cluster about the origin. However, at the origin itself the power of the imagination seems to leave us in the lurch. If the author of this book were involved in a discussion, he would side with Berkeley: "If any man has the faculty of framing in his mind such an idea . . . it is in vain to pretend to dispute him out of it, nor would I go about it. All I desire is that the reader would fully and certainly inform himself whether he has such an idea or no."[2]

Now, does the curve represent a continuously connected line at the origin? We first ask what characterizes a continuous connection. Let the reader draw a portion of a curve. Then if he selects a point with the coordinates x, y and changes x just a little, he will note that y, too, changes only insignificantly. To be more precise, a curve is continuous if to *sufficiently* small changes of x there correspond arbitrarily small changes of y. If the reader looks at the curve $y = \sin \frac{1}{x}$ from this

---

[2] Translator's note: *A Treatise Concerning the Principles of Human Knowledge*, The Works of George Berkeley, Vol. I, p. 146.

12. REMARKABLE CURVES

standpoint, especially at the origin, he notices at once the discontinuous nature. For, within an interval, no matter how small, which has 0 as an endpoint (say from 0 to ε), the curve will always oscillate between $+1$ and -1; shrinking the interval on the x-axis will not force the variation of the function to become less, for example, than the value $1/10$.

Let us now modify this example by considering the function $y = x \cdot \sin \frac{1}{x}$. The presence of the factor x causes the amplitudes of the oscillations themselves to change with x; thus they are large if x is large, and small if x is small. As a consequence the wave trains actually cluster infinitely densely about 0, but at the same time lose in altitude so that somehow they fade away as an infinitely fine ripple on the beach. The entire curve is now enclosed between the two straight lines $y = x$ and $y = -x$, and has the following appearance:[3]

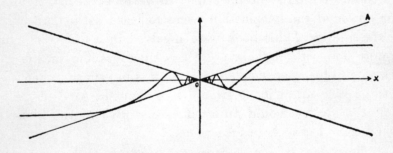

Figure 18

This curve is continuous at the origin. For to a small change of x there also corresponds a small change of y. What direction does the curve have at the point $x = 0$? By the direction we mean, of course, the limiting position which the secant approaches if the point P is held fixed while P_1 moves along the curve closer and closer to P. Now, as a point moves along the wave

---

[3] Cf. footnote[1].

12. REMARKABLE CURVES

curve toward 0, it will be all the way up infinitely often and all the way down infinitely often (lie on OA, respectively, OB). Consequently, the secant constantly moves to and fro within the angle AOB without deciding upon a definite limiting position. The only conclusion which can be drawn from this is that the curve does not have a tangent at the point 0; this means that it is directionless even though its course is completely continuous.

This example can indeed be used to study the fact that the value of a differential quotient depends entirely on *how* Δx tends to 0. For example, if Δx runs through the sequence $\frac{2}{\pi}, \frac{2}{5\pi}, \frac{2}{9\pi}, \frac{2}{13\pi}, \ldots$, then all difference quotients are 1 and therefore the differential quotient is also 1. If Δx runs through the sequence $\frac{2}{3\pi}, \frac{2}{7\pi}, \frac{2}{11\pi}, \frac{2}{15\pi}, \ldots$, then the difference quotients are -1 and the differential quotient is -1; if Δx runs through the sequence $\frac{1}{\pi}, \frac{1}{2\pi}, \frac{1}{3\pi}, \frac{1}{4\pi}, \ldots$, then all difference quotients are 0 and the differential quotient is also 0. By choosing a suitable null sequence the differential quotient can be assigned any arbitrary value between $+1$ and -1.

One was inclined to hold this as an exceptional phenomenon which could occur at the most at individual points. Then Weierstrass amazed the mathematical world with the discovery that there are curves which are continuous everywhere but differentiable nowhere, namely, which behave at *every* point just as our curve does at the origin. Obviously such curves no longer correspond to the image associated with this word.

The idea of Weierstrass rests pretty much on the process of superimposing on a wave curve a finer one, on this a still finer one, etc., ad infinitum, so that finally a curve is generated which never runs "smooth" but is curled and ruffled up to infinity, just as the curve $y = x \cdot \sin \frac{1}{x}$ at the point $x = 0$.

12. Remarkable Curves

The development of Weierstrass would be too troublesome to set forth. However, today we are fortunately able by much simpler methods to construct curves which behave like the one discovered by Weierstrass. A very simple example is due to H. v. Koch.

We start with the unit segment and delete the middle third with the exception of the endpoints which are left standing. We replace the gap by a peak by constructing the two sides of an equilateral triangle over the deleted third. We continue in this way with each of the line segments which remain intact, namely, we again delete the middle third and replace it by an appropriately smaller peak. The accompanying figure shows the first stages of the construction. If this process is continued without end, a curve is generated which is continuous and nowhere differentiable.

That the curve is continuous rests on the fact that the size of the piled-up peaks becomes smaller and smaller. That it is not differentiable will first be made evident for the point A. If we follow the path of a point moving from A, let us say, along the approximating polygon last drawn, it will climb up many times to a summit (as in P_1 and P_3) and dip down many times to the x-axis (as in P_2 and P_4). Now the reader must imagine that the situation which he sees in the large in this figure repeats itself over every small portion as the process is continued, so that every segment, no matter how small, is replaced by a similar, though appropriately smaller, image of the whole figure. Therefore, if the point moves along Koch's curve toward

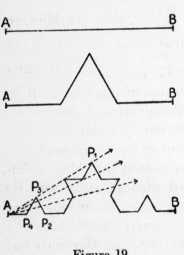

Figure 19

12. REMARKABLE CURVES

A, the secants passing through A will move up and down infinitely often, running back and forth between 60° and 0° without approaching a definite limiting position. Perhaps the reader has the impression that in the immediate neighborhood of A the curve must run horizontally for a little way. This is an illusion. For the peak pattern of the whole figure rises again over every segment, no matter how small, which begins at A. This means that every neighborhood of A, no matter how small, contains points of the curve distributed in an angular sector of 60°. Entirely analogous considerations can be applied to any other point of the curve.

Up to this time, mathematics had failed to realize that such relations were possible. It was taken for granted that a continuous function was also differentiable.[4] Therefore, the discovery of Weierstrass deeply undermined the naive trust in intuition. This shock stirred up the general question: what, strictly speaking, is a curve? This question is one of those that are easier to ask than to answer. We will try to give the reader an insight into the difficulties encountered in setting up a precise definition.

About 1880, C. Jordan set up a definition which renders approximately our intuitive conception of a curve. This definition says basically the following: a curve is generated if a point runs along in continuous motion. The motion of a point is controlled if we can state at each instant of time where the point is, that is, if we know the position as a function of time. If the point moves in a plane, the position is determined by two numbers x, y, the time by one number t. Hence the motion will be completely described by stating how x and y depend on t, that is, by giving x and y as functions of t

---

[4] There still prevailed in the mathematics of those days, as F. Klein put it, the "paradisiacal" state in which one did not distinguish in the case of a continuous function between good and bad, differentiable and non-differentiable.

12. Remarkable Curves

$$x = \varphi(t)$$
$$y = \psi(t).$$

Besides, the functions φ and ψ shall be *single-valued*, which means that at every instant the abscissa and ordinate of the position of the particle shall be uniquely determined. We speak of a *continuous* motion if the functions and are continuous. For such a motion the moving particle occupies arbitrarily adjacent points of space at instants which are sufficiently close together; in other words, sudden bounds (vanishing at one position and reappearance at another) are excluded. If we assume that the motion takes place in a unit interval of time (let us say, in a minute), then we can restrict t to the interval from 0 to 1. Jordan's definition then amounts to the following: *a curve is a continuous and single-valued image of the unit segment*. On the other hand, we must be resigned to the fact that a circle, for example, would then no longer be a curve, since it runs back on itself whereas the unit segment has two ends. Except for the latter condition, this definition seems to agree exactly with what we have in mind when we use the word "curve." All the more remarkable is a discovery of Peano. In 1890 he showed that there are structures which are single-valued and continuous images of the unit segment, and so are curves in the sense of Jordan, but which, in the strict sense of the word, fill up an entire square. If this curve were to be drawn, the entire surface of a square must be colored black—this would be the picture of the curve.

We will now give a representation of this remarkable curve in a simplified form which is due to Hilbert. The problem which Peano and Hilbert had set up was the following: given the unit segment and the unit square, how can these two structures be related to each other, so that to every point of the unit segment there corresponds exactly one point of the square, and so that the correspondence is continuous? To illustrate this by a graphic example, let us think of a traveler who shall pass through all points of a square in one minute. What will his travel route look like?

We divide both the minute and the square into four equal parts and arrange it so that the traveler always moves through exactly one-quarter of the square during a quarter of a minute. In the first quarter minute he is to wander through the first quarter of the square without touching it again later; in the second quarter minute, the second quarter of the square, etc. The reader will say that this has not simplified the problem. For, how shall the traveler manage to pass through a whole surface in a definite time? Aren't we confronted with exactly the same difficulty as at the beginning? No! The following procedure could be headed by the motto: "If you cannot do it in the large, then begin in the small." We continue the division process; namely, we divide each quarter of the surface of the square again into four equal parts, so that the entire surface of the square is divided into 16 small squares. Let us number these 1, 2, ... 16. We correspondingly divide the minute into 16 equal intervals. These, too, are numbered 1 to 16, and to each subsegment we associate the subsquare with the same numeral. In so doing we must take care that two squares with successive numerals have a side in common. By the continued repetition of this process the division of the square and the segment can be arbitrarily refined. In passing on to the limit the distinction of dimension vanishes, for a point is obtained as a limit no matter whether we started from the division of a line segment or from the division of an area.

This makes it possible to map segments and squares on one another in the desired manner. For this purpose we have to prove three things. 1. To every point of the unit segment there corresponds a point of the square (which means that at every instant the traveler finds himself at exactly one position of the surface). 2. No point of the square is left out (the traveler gets everywhere). 3. The mapping is continuous (he moves on an uninterrupted line).

161

12. REMARKABLE CURVES

As to 1. If we select some point t on the unit segment, then as the division is continued only two cases can occur.

a) The point t always falls inside a subsegment (for example, $t = \frac{1}{3}$). In this case let us think of the consecutive intervals in which t falls at the successive steps of division as marked. These will become shorter and shorter and contract to the point t. Now our procedure associates a subsquare to every such interval. These subsquares form a never-ending sequence of parcels, which, thanks to our division rule, are nested in one another and contract to a definite point q. Consequently, to the point t there is associated the point q.

b) t coincides after some steps with a division point (for example, $t = \frac{3}{16}$). In this case we combine the two subintervals which meet at t and proceed as before. The only difference is that double squares correspond to double intervals which naturally again converge to a definite point q. In any case, to a point of the unit segment there corresponds one and only one point in the square.

2. We will now select some point out of the area of the square and prove that there is at least one point of the unit segment which corresponds to it. This will show that our correspondence overlooks no point of the square. We again distinguish two cases.

a) The point q always lies inside a subsquare, no matter how far the division is continued. q then determines a sequence of nested squares; to these there corresponds a sequence of nested intervals which contract to a definite point t.

b) q falls on the boundary or at the corner of a square where two or more squares with non-successive numerals meet. In this case, we can advance toward this point from various sides. If the reader makes a sketch of a square which is divided into 16 parts, as in the figure b), p. 164, then he will note that three squares with non-successive numerals come together the center. Consequently, convergent processes can be launch from three different sides which all tend to one and the same point. To these there evidently correspond on the unit segment three entirely separate nests of intervals which there determine the three points t_1, t_2, t_3.

Hence the correspondence is not unique in the reverse direction. In the above case there are three distinct instants at which the traveler arrives at the same point of the square. This means that the path of the curve intersects itself.

As to 3. The path of the route never breaks off. This, strictly speaking, follows immediately as soon as we consider more closely the nature of the correspondence. A break in the path of the curve could only occur if the point jumps, that is, if the point is at wholly different positions of the surface at successive instants. However, this is impossible. Now, let the reader visualize two points on the time axis very close to each other. If he thinks of the unit segment as divided into an appropriately large number of intervals, these two points will fall in the same interval or in two adjoining ones. But then this will also be true of their images in the square, that is, these too will fall in the same small subsquare or in two adjacent ones. The closer the points lie on the time axis the more will their images be drawn closer in the square. Nearness in time will therefore correspond to nearness in space.

If t runs through the unit segment, the image point q describes a continuous linear path filling up the entire square. To illustrate Peano's curve we may use the following approximating polygons:

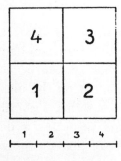

a

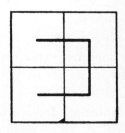

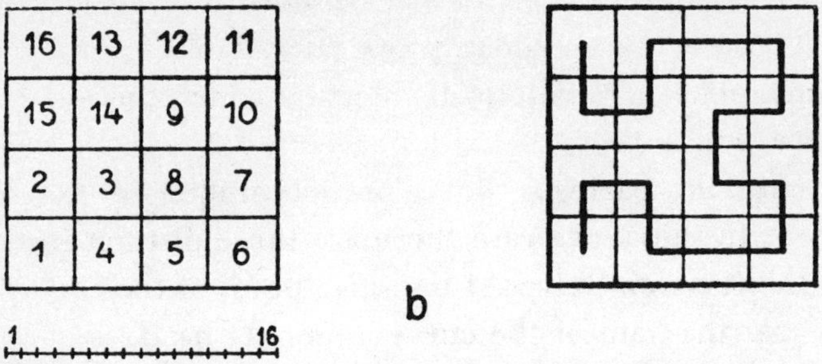

12. REMARKABLE CURVES

Figure 20

The polygons indicate the sequential order in which the individual subsquares are traversed by the curve. We have to visualize that every rectilinear segment will be replaced at the next step of the approximation by a broken line path. Peano's curve is the infinitely distant limit structure which the individual meandrous paths approach.

Is this curve differentiable? By the tangent, we remember, we agreed to understand the limiting position of the secants. Let us now think of a point P in the area of the square as marked and connected with another point P_1. What happens if P_1 draws near to P along the curve? This can be realized by bringing t_1 closer and closer to t; as t_1 moves toward t, that is, passes through an interval, the associated image point runs through a whole portion of the area. Therefore, if I connect P with P_1, the secant ray will move to and fro without approaching a limiting position. In fact, if the neighborhood of t is suitably chosen, it can be made to point in any direction of the compass.

This discovery is admittedly somewhat paradoxical, and it is sometimes explained as if to imply that intuition had been caught in an error. In this sense one has spoken of a crisis of intuition. Hence we will spend a moment on the question: does intuition deceive us if it seems to say that a continuous line which runs without corner points possesses a definite direction at every point? I do not believe that the preceding

12. Remarkable Curves

discussion justifies this conclusion. This will become clearer if we consider how the concepts of the mathematicians are related to the intuitable structures familiar to us.

For example, what is an intuitively evident curve? Strictly speaking, three types of things are designated by this word: a stroke whose width is negligible relative to its length, the boundary between two colored plane surfaces, and a certain kind of motion ("curves follow a course"). At any rate, we get to know the meaning of this word through a demonstrative definition, just as in the case of the words "stroke," "wire," "thread," etc. This means that we call anything a "curve" which is somewhat similar to a paradigm (say, to a drawn crooked line) without specifying precisely the kind of similarity; therefore, this concept is not sharply delineated. Consequently, there will be cases where one will waver between calling something a curve or not. The concept is as diffusely defined as, for instance, the concepts "longish," "roundlike," "notched," "frayed," etc. In applying these words certain zones of indefiniteness are encountered, which we can designate as "suspended classes." The concepts of mathematics, however, are precisely defined.

How can we attain such a precise definition? The procedure is very clearly described in the following excerpt from Menger's *Dimensionstheorie* (Theory of Dimension), p. 75f.:

"If a word which already has a meaning attached to it in daily life is to be precisely defined in science, there is no reason for setting it in contradiction to the daily usage of the word, that is, for excluding from the concept things generally designated by the word under consideration, or including in it things generally not designated by it. Hence a *formal requirement* for the rigorous definition of a word appearing in the colloquial language is the following: it should *make precise and complement the usage of the word* which is vague and incomplete in border cases *without contradicting the same.*

"There are various ways, however, of attaining this sort of complementation and preciseness; indeed, it can follow in

12. REMARKABLE CURVES

various ways not only formally but also (with regard to the suspended classes) as to content. For example, certain structures have been developed of which an individual does not know whether they are one-dimensional, or on which different men give different answers. Hence such structures can be taken as one-dimensional just as well as not, without the definition, on this account, contradicting the general usage of language. Every attempt to attain a definite preciseness contains, therefore, a certain amount of *arbitrariness* that can only be justified by the *fruitfulness* of the definition. The purpose of words in daily life is understanding among men; the purpose of a rigorous definition is to form the starting point of a deductive system."

If this is kept in mind, it may very well change our attitude toward the question whether intuition deceives us. The mathematician does not wish at all to describe the relations submitted by intuition. He uses a system of concepts which agrees with intuition only here and there and then deviates from it; consequently, we should not be surprised if discrepancies arise. But then it is unreasonable to blame intuition; instead we should strive to attain a better understanding of the intuitive concepts in their individuality. (Regarding this point, cf. the end of Chapter 13.)

The discovery of Peano seems to obliterate one of the most fundamental distinctions, namely, the distinction between dimensions. If a curve can fill up a plane surface, how shall we then distinguish between the one- and two-dimensional entities? This leads us to a deeper-lying question, which we will discuss here at least to the extent of touching on the detail.

First of all, we must present a remarkable discovery of Cantor's, namely, the fact that a line segment can be mapped uniquely on a square, that, therefore, a surface can be made out of a line segment by merely rearranging its points. To express ourselves more precisely, we take the unit line segment and a square whose sides have length 1, and now show that t two structures can be uniquely related to one another, so that to every point of the line segment there corresponds exactly one point of the square and to every point of the square exactly one point of the line segment.

To furnish this proof we first recall that all real numbers can be written in the form of decimal fractions, and indeed as *infinite* decimal fractions. Thus, for example, $\frac{1}{2} = 0.5 = 0.4999\ldots$ We think of all decimal fractions between 0 and 1 as represented in this form.

In order to determine a point of the square two numbers must be specified; the simplest method is to use its distance from two adjacent sides of the square; these we call x and y. Since the point lies inside or on the edge of the square, the numbers x and y only lie between 0 and 1; therefore they can again be written as decimal fractions, and indeed as non-terminating decimal fractions. Hence to each point of the square there uniquely corresponds a certain pair of decimal fractions, and the set of all points of the square is represented as the set of all possible pairs of decimal fractions x, y.

The required mapping will now amount to associating a certain pair of decimal fractions x, y to every decimal fraction t, and conversely. This is made possible by the simple idea of "splitting" a given decimal fraction into two new decimal fractions such that the given decimal can again be found by inverting the procedure. In order to carry this out further, let us assume that the decimal fraction

$$t = 0 \cdot a_1 b_1 a_2 b_2 a_3 b_3 \ldots$$

is given. Out of the odd places, respectively, the even places, let us form the decimal fractions

$$x = 0 \cdot a_1 a_2 a_3 \ldots, \quad y = 0 \cdot b_1 b_2 b_3 \ldots$$

whereby the decimal fraction t — or rather its stock of numerals — is, so to speak, separated into two new decimal fractions, and this separation is possible in only one way. The number

12. REMARKABLE CURVES

$t = 4/11 = 0.363636\ldots$ leads, for example, to the separation $x = 0.333\ldots$, $y = 0.666\ldots$, which means that the number $t = 4/11$ corresponds to the pair of numbers $x = 1/3$, $y = 2/3$. Now if x, y is interpreted as a point in the surface of the square, then this rule associates exactly one point of the square to every point of the unit segment.

Conversely, to every point of the square there is also associated one point of the unit segment. Thus, let us select an arbitrary point from the surface of the square. We first determine its two distances x, y from the sides of the square. As these distances lie between 0 and 1, we can represent them as decimal fractions, and actually as non-terminating decimal fractions:

$$x = 0 \cdot a_1 a_2 a_3 \ldots, \quad y = 0 \cdot b_1 b_2 b_3 \ldots$$

Next, let us merge these two representations by compounding the new decimal fraction

$$0 \cdot a_1 b_1 a_2 b_2 a_3 b_3 \ldots$$

out of its numerals. It is obvious that this is exactly the same decimal fraction whose separation leads back again to x and y, that is, it is the decimal fraction t. In virtue of our procedure, therefore, a point of the square has actually been assigned to every point of the unit segment and conversely.

To this discussion an objection can be raised which we will illustrate by an example. What point of the square corresponds to the point

$$t = 0.33030303\ldots ?$$

Evidently the point

$$x = 0.3000\ldots$$
$$y = 0.3333\ldots$$

However, the difficulty with this is that a number of the form is forbidden. If we exclude the terminating decimal fractions, then we no longer obtain all possible values of t; as a point

168

passes through the square, its image point does not run through the whole unit segment. However, this difficulty can be removed, according to König, by interpreting $a_1, a_2 \ldots b_1, b_2, \ldots$ not as individual digits but as certain complexes of digits — they might be called the "molecules" of the decimal fraction — by always including with a digit of the decimal fraction, different from zero, all the zeros which immediately precede it. Thus, if we group the digits of t in the following manner

$$t = 0 \cdot 3 \,|\, 3 \,|\, 03 \,|\, 03 \,|\, 03 \,|\, \ldots$$

then the partitioning process gives

$$x = 0.30303 \ldots$$
$$y = 0.30303 \ldots$$

Now *every* number t can be split up in the described manner without giving rise to terminating decimal fractions. This means that line segments and squares can be put into a **one-to-one correspondence**.

Cantor's discovery seems paradoxical, for it had always been believed up to then that a plane surface must contain infinitely many more points than a straight line. Since the reader is perhaps not satisfied with the proof and may have the feeling that our argument contains some trick which he has not seen through (Schopenhauer once expressed this distrust with the words: "That this is so, we must grant, coerced by the law of contradiction: why it is so, however, we are not told. Hence we have almost the unpleasant sensation as after a trick of jugglery. Often an apagogic proof closes all doors one after another and leaves only the one open through which we must now enter merely for this reason.") we wish to say a word or two about this matter. The startling fact is that the proof shows us that a square can be made out of the points of a line segment, even though the points of the line segment form only an infinitely small part of the points of the square. Shall we trust a proof which leads to such abstruse consequences?

12. Remarkable Curves

Let us give the matter another thought. It is quite true that the part is smaller than the whole *as long as we restrict ourselves to finite sets*. However, for the present, it has not yet been established that this proposition is also valid for infinite sets, for we are here passing over to an entirely new domain. Actually we do not know at all what we are to understand by "equal," "greater," "smaller." These concepts must perhaps be defined anew, and if this is done it may turn out that the old laws for the relation of the whole to the part can no longer be maintained. The infinite sets just have to differ somehow from the finite sets, and for this very reason we must not expect all properties of the latter to be found indiscriminately in the former. After due consideration we lose our astonishment that the time-honored proposition "The part is smaller than the whole" is no longer valid for infinite sets. Indeed, infinite sets are actually characterized by the fact that they violate this law. Dedekind came thus to the ingenious idea of defining infinity just by this paradoxical property. From this standpoint a system of things is infinite if it can be related term for term to a part of itself. We comprehend the progress of this definition by saying that whereas all earlier explanations of infinity always conceived of it only in a negative way, as of the non-finite, here infinity is characterized for the first time by a positive inner property.

Now this should be enough regarding infinite sets. We next ask: Is the correspondence discovered by Cantor continuous? We will immediately find the answer to this question on investigating the behavior of the correspondence, let us say at the point $t = \frac{1}{2}$. If t is written in the form

$$t = 0.49999\ldots,$$

then the separation which was described earlier yields

$$x = 0.499\ldots$$
$$y = 0.999\ldots,$$

which means that to the point $t = 1/2$ there is associated the point q_1 in the square with the coordinates $(1/2, 1)$. Now what happens if t becomes just a little larger than this value? Let us assume that t is $0.50000111\ldots$. Then

$$x = 0.500111\ldots$$
$$y = 0.00111\ldots$$

Hence as t passes through the critical value $1/2$, the image point makes a bound from the upper edge of the square to the lower (in the figure from q_1 to q_2). Such bounds are found whenever t is a decimal fraction in which all digits are nines after a while, namely, whenever t can also be written as a terminating decimal fraction, and these numbers are densely sown. If t stands for time, then the point q does not move continuously at all; instead it flashes through the square by jerks, if one is permitted to use this expression.

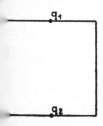

Figure 21

Let us now compare the correspondences of Peano and Cantor. Peano's correspondence is

1. single-valued in the direction from the line segment to the square (this means that to a point of the unit segment there corresponds exactly one point in the square);

2. many-valued in the direction from the square to the unit segment (this means that to a point of the square there correspond in general several points of the unit segment);

3. continuous.

Cantor's correspondence is

1. single-valued in the direction from the segment to the square,

2. single-valued in the direction from the square to the segment, therefore, it is one-to-one, but

3. discontinuous.

Consequently, if the correspondence is continuous, it is not one-to-one; if it is one-to-one, it is not continuous. Is this per-

171

haps due only to chance? Isn't it possible to invent a correspondence which has both properties, namely, is one-to-one and continuous? Now the remarkable fact is that this cannot be done at all. Jürgens has discovered a very simple proof of this, which we will give here. Let us assume that there is a correspondence of the line segment to the square which is one-to-one and continuous; we will show that this assumption leads to a contradiction. Thus let us select two points q_1, q_2 out of the area of the square. By assumption there are two points t_1 and t_2 on the unit segment which correspond to them. Now, if a point moves continuously on the area of the square along some path from q_1 to q_2, for example, along the path I, then by assumption the associated point must also travel along the unit segment continuously from t_1 to t_2, namely, runs through the entire interval between these two points. But what if the point of the square had followed a route other than the one it had chosen, but which also leads from q_1 to q_2, for example, the route II? Then the associated point must likewise run through the entire interval from t_1 to t_2. This interval would therefore be mapped on two distinct portions of lines of the square, contrary to the assumption that a point of the line segment is to correspond to only *one* point in the square. Hence, if we wish to carry continuity through, the one-to-oneness must be sacrificed.

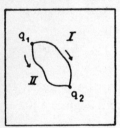

Figure 22

The famous proposition of the invariance of dimension, which was first proved by L. E. J. Brouwer (1911), represents the last word in this investigation. By this proposition a portion of a k-dimensional continuum can only be put into one-to-one and continuous correspondence to another portion of a k-dimensional continuum, never to a portion of a m-dimensional one if $m \neq k$.

We started with the question: how shall we formulate conceptually the distinction between curves and surfaces? T

12. REMARKABLE CURVES

answer given by authors as recent as Riemann and Helmholtz, namely, that a point on a curve is characterized by one coordinate, a point on a surface by two coordinates, does not completely answer the question. For today one must say that it is arbitrary how many coordinates one wishes to use. If we are only interested in distinguishing the points of a square from one another, namely, in providing them with proper names, then a single real coordinate will do. Evidently, what Riemann and Helmholtz had in mind is that the distribution of the numbers over the points shall be *continuous*: then, in the case of a surface two coordinates are actually needed, no more and no less.

However this has still not given us an answer to the question: "What is a curve?" For example, to say that a curve is a single-valued and continuous image of the unit segment, would mean to exclude those curves which run back into themselves, those which intersect themselves, and those which branch out; furthermore, also structures as that represented in Fig. 17, to which everybody will concede the property of linearity. The problem of characterizing in general the concept of curve was definitely solved by Menger and Urysohn, after an antecedent history tied to the names of Poincaré and Brouwer. In a few words the solution can be indicated as follows. The most characteristic property of curves is their one-dimensionality. The problem requires the explanation of the concept of dimension number. Let us think of some geometrical structure (a point set M) in our Euclidean space. It can have at various points a different dimension number; for example, surfaces may grow out of a solid as leaves from a stem, and curves may go out of these as thorns. We will now try to explain what it means to say that such a structure has the dimension number 1, 2, or 3 at a definite point. For this purpose Menger[5] employs the following mental experiment: how can a point with its neighborhood incident in M be detached from the remaining space? If the point lies inside a three-dimensional structure (say, a solid piece of

[5] *Dimensionstheorie* (Theory of Dimension), p. 78.

12. Remarkable Curves

wood), then the rest of the solid is removed with a saw and therefore a whole surface must be sawed through; if it lies on a two-dimensional domain (say, on the surface of a leaf), then the neighborhood is detached with scissors by cutting through a curve; if it lies on a one-dimensional structure (a fine thorn), then the thorn has to be pinched through at only two points (or, at any rate, at finitely many). But this means that a set is three-dimensional at a point if within the set there are arbitrarily small neighborhoods of this point whose boundaries are two-dimensional; a set is two-dimensional at a point, if inside the set there are arbitrarily small neighborhoods of this point whose boundaries are one-dimensional; a set is one-dimensional at a point, if inside the set there are arbitrarily small neighborhoods of this point whose boundaries are zero-dimensional (in ordinary language, which consists of single points); a set is zero-dimensional at a point if in the set there are arbitrarily small neighborhoods of the point whose boundaries are (-1)-dimensional; *and the empty set shall be the only (-1)-dimensional set*, which means, the set that contains no element. This shows how the dimension number can be defined by a recursive process from an absolute starting point—the concept of dimension (-1).

These ideas are typical of a specific class of geometrical investigations in which one is not concerned with the study of lengths and angles, nor with facts connected with the concepts of straight or crooked, but only with properties which remain unchanged by continuous one-to-one transformations. The reader may think of a figure drawn on a rubber surface, this surface being arbitrarily distorted without tearing up or new contacts appearing. All lengths and angles will thereby be changed; a straight line may be distorted into a curve as scrawly as one wishes, and it still turns out that other definite properties remain wholly unchanged by this distortion, for example, the dimension number. In topology one studies precisely those properties of a spatial structure which remain invariant under arbitrary distortions of this kind. However, this leads us over to a new train of thought.

Appendix: What Is Geometry?

If the layman were asked what he understands by geometry, he would perhaps say that it is the science of the properties of space structures, that is, of geometrical properties. If he were now called upon to define the concept of geometrical property more precisely, he would note that his answer was by no means sufficient, and that just at this point a new question arises. Namely, is every spacelike property also a geometrical one? We shall see! If I find, let us say, by measuring a segment that it is ½m. long, then this is certainly a property of this segment; however, geometry is not concerned with the investigation of such properties. F. Klein once illustrated this point very drastically through the statement of a colleague who had asserted: "If the center of the inscribed circle and the center of the circumscribed circle of a triangle are marked, then the second always lies 3 mm. east of the first"; repeated measurements had convinced him of this. If such a proposition were true, then it would describe a fact of topography or geography. On the other hand, a theorem about the equilateral triangle is a proposition of an entirely different sort, since it expresses a fact which not only occurs this once but always, i.e, it is true not only for one configuration but for a class of configurations. If we ask what is common to the configurations of a class, the answer is: their similarity. First, we disregard the position the configuration occupies in space; secondly, its size. This means that we look at only those properties which the configuration shares

12. REMARKABLE CURVES

with all others of the class. Whatever does not belong to all configurations of a class, merits only an individual interest and is excluded from geometry.

What is to be understood by a class of configurations depends on us, and in this train of thought we are readily confronted with the problem of distinguishing various "geometries," each according to the partitions agreed upon. In our example all configurations of a class are generated if one of these is selected and submitted to two kinds of transformations, namely, an enlargement or reduction with preservation of form. (In the case of spatial structures, mirror reflections are also to be taken into consideration which, for example, bring a right hand glove over to the left hand.) Now, from one configuration new ones can be derived in many other ways, for instance, by central projection. If we photograph a configuration drawn on the blackboard from the side, different pictures can be obtained which in general are no longer similar. If we now think of projecting the configurations in all conceivable ways, a new class of configurations is generated, more extensive than the earlier one, and we are confronted with the problem of determining whether there are properties common to all configurations of such a class. It is easy to see that there actually are such properties. Let the reader draw any two straight lines and choose three points on each; the two sequences may be designated by A, B, C, respectively, A', B', C'. If he now connects the points crosswise (this means, A with B', A' with B; A with C', A' with C; B with C', B' with C), and marks the three intersection points, he will find that they lie on a straight line. This property is preserved if the entire configuration is photographed. Obviously the formulation

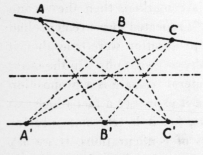

Figure 23

of this proposition cannot be made to depend on lengths or angles in the configuration, but only on the fact that certain points lie on a straight line. *Projective Geometry* is concerned with the investigation of these properties. From this standpoint the proposition concerning the equilateral triangle is no longer spoken of as a geometric property; indeed, from the projective point of view it is actually no longer possible to distinguish the equilateral triangle from the other triangles. The properties which projective geometry brings to light are more intimately tied to the space structures and harder to destroy than the properties considered in the usual geometry.

If we had chosen transformations by parallel projection instead of by photography, then we would encounter another geometry. It would occupy a middle position between the usual and the projective geometry and is called *affine geometry*. Let the reader think of two arbitrary planes in space and a configuration be drawn on the first plane. What properties of this configuration remain unchanged if the configuration is transferred to the second plane by parallel rays? If the configuration is a circle, it will go over into an ellipse. Circles and ellipses are no longer distinguishable from the standpoint of affine geometry. Projective geometry goes still further by bringing parabolas and hyperbolas in the same class with these two curves, for they can all be carried into one another by projection. Hence, on ascending from the usual or "metric" geometry to affine and then to projective geometry, the differences of the configurations vanish more and more, and the deeper lying, more general, characteristic features come to the fore instead. It is as if one were stepping further and further away from an object: the details flow together and we see only the large lines.

If this concept is to be more sharply delineated, we must have recourse to the fundamental concept of *group*. This is the name given by the mathematician to certain systems of operations which are closed in themselves. Let us consider, as an illustration, a sphere into which a tetrahedron is inscribed,

12. REMARKABLE CURVES

and let us fix our attention on those rotations of the sphere which take the tetrahedron into itself. The reader can easily establish that there are exactly twelve such rotations. If two such rotations are carried out one after another, a rotation of the same kind is again obtained. The successive execution of two rotations will be called their composition. Our twelve rotations now have the property that the composition of any two is again one of these rotations, that is, it will never lead us out of the domain of these twelve operations. This is the most important property of a group. The mathematicians encounter such remarkable structures in various sorts of investigations and have gradually sifted out their properties. The complete definition says: a group in a system of entities—for instance, operations—finite or infinite in number, which satisfy the following five conditions:

1. There shall exist a rule by which two entities can be composed with one another. The entities are called the elements of the group and are designated by A, B, etc. The composition, after the manner of multiplication, is denoted by $A \cdot B$. The first condition now says that the composition $A \cdot B$ shall be single-valued.

2. The element $A \cdot B$ shall again belong to the same system; therefore, one shall not be able to step out of the system no matter how the elements are composed.

3. The composition shall be associative, which means that $A \cdot (B \cdot C) = (A \cdot B) \cdot C$. Commutativity, on the other hand, will not be required.

4. There shall exist in our system a definite element E which, if composed with any other element, again renders this element, so that this element satisfies the relations: $A \cdot E = E \cdot A = A$. (The element E is called the "unit element," since with respect to the composition of elements, it plays a role similar to that of the number 1 with respect to the multiplication of numbers.)

12. REMARKABLE CURVES

5. To every element $\bar{A}$ there shall exist an inverse element t such that

$$A \cdot \bar{A} = \bar{A} \cdot A = E$$

An example of a group is the totality of motions in space. Thus, the composition of two motions is again a motion and every motion can be cancelled by an inverse one; the unit element corresponds in this example to rest as the limiting case of motion.

On reconsidering the various geometries, we find that the spatial transformations which produce a definite class of configurations always exactly form a group. Thus, the transformations which leave the elementary-geometric properties of a structure unchanged are the motions, the similarity transformations and the process of reflection. These transformations (along with all of their compositions) form a group which is designated as the *principal group* of the spatial transformations. We can now say that ordinary geometry deals with those properties which remain invariant relative to all transformations of the principal group. Anything that heretofore has instinctively been considered as an "essential" property of a geometrical structure, as opposed to an "incidental" one, has thereby gained an exact formulation.

If the principal group is replaced by a more comprehensive one, then only some of the properties remain invariant. If the parallel projections are added to the above, the *affine group* is generated; if in addition the central projections are included, the *projective group* is generated; these groups play the very same role in affine, respectively, projective geometry as the principal group in "metric" geometry. The more comprehensive the group is which we take up, the more we penetrate into the depths. The properties which metric geometry investigates are the easiest to destroy; they lie, so to speak, at the uppermost stratum. The affine properties lie somewhat deeper; the pro-

12. REMARKABLE CURVES

jective, still deeper. In this interpretation the group appears as the guiding principle, that is, a type of geometrical investigation arises only if a manifold is given along with a group of transformations in this manifold. We can say with Felix Klein, who first pursued these ideas in his "Erlanger Programm" that *Every geometry is a theory of invariants relative to a definite group.*

We are here presented with a view of an infinite series of geometries which arise by extending the principal group in various directions. Thus, we can obtain the "geometry of inversion" by starting from the metric geometry. This geometry comprehends only those properties of a configuration which remain unchanged not only under the transformations of the principal group but under reflections with respect to a given circle (the inversion transformation). It can be shown that these transformations convert, for instance, straight lines and circles into one another, so that these two kinds of configurations must be included in one class.

Finally, we take a last step and reach for the geometry of all one-to-one and continuous point transformations. Here we seek those properties which are invariant under arbitrary continuous distortions. From this standpoint, for example, a sphere, a cube, and a pyramid are not essentially distinct; each of these structures can be converted into one another by continuous distortions. On the contrary, for example, a sphere and a torus (an air tube) belong to entirely distinct classes of solids, since it is obviously impossible to convert one structure into the other continuously by a series of intermediate forms. The properties which now come to light form the subject matter of *topology*. For example, there is a theorem in topology which says that knots are possible neither in two- nor in four-dimensional space, but only in a three-dimensional space. Another example of a topological property can be made clear on the Möbius strip. Let the reader take a strip of paper and bend its two ends together, so that a closed ring is formed; let, however one end be twisted beforehand through 180°, so that the front of the strip goes over into the back—he then obtains a surface which has *only one side*. A person moving on this surface would actually return to the same place after a revolution, but would now find himself on the opposite side of the strip of paper; therefore, it no longer makes sense to distinguish the two sides of the surface. It is clear that this property remains invariant relative to arbitrary continuous transformations and that the distinction between one-sided and two-sided surfaces has a topological significance. In order to state a problem in this domain, too, I will cite the "four-color theorem": Geographers have found empirically that no matter what the political map of a continent looks like four colors are sufficient to differentiate the various countries. Up to now, an exact proof of this theorem has not been obtained.

On going still further to the most general one-to-one point transformations, we arrive at the standpoint of the *theory of sets*. From this point of view the distinction of dimensions is also vague: a line segment, a surface, a solid can be mapped into one another point for point, that is, they are only different representatives of one class. Of the whole rich world of geometrical forms there now remain only very few traits, such as the distinction between finite and infinite point sets.

The concept of dimension belongs to the system of topological concepts, and we now clearly recognize why it has no place in the most general investigations of Cantor aimed at the one-to-one transformations.

Hence in the series of geometries, metrical geometry stands at one end (or, strictly speaking, topography which regards every geometrical structure as an individual); at the other end, the theory of sets. Any other geometry holds a definite place between these extremes.

13. The Real Numbers

Now and then popular publications appear in which the question is raised whether the irrational numbers really exist; these are an echo of an earlier period of science in which the nature of these numbers was still immersed in a mysterious darkness. As a matter of fact, the reason for such uncertainties can be readily understood. Discussions like those employed in the previous chapters are not easy to understand. We spoke there, for example, of nests of intervals which contract to a point. Strictly speaking, such an expression makes a strong demand on our thinking. If I cut out of a line segment another, from the latter again another, etc., then I certainly always obtain a *line segment*. It can by no means be found out how a line segment is to turn into a point. The gap between the two always seems to remain the same. "In the large or in the small," says P. du Bois Reymond,[1] "the line segment always remains a line segment between two rational points. Now, if we suddenly allow, without logical foundation, a *point* to replace a *line segment*, this is an act by which we quite arbitrarily introduce a new idea, let it abruptly follow the first, and assume beforehand exactly what should be proved. A *gradual* coincidence of two points, as this event is sometimes described, is all the more meaningless. The points are either separated by a line

---

[1] *Die allgemeine Funktionentheorie* (*General Theory of Functions*) 1882, p. 61.

segment, or there is only one point; there is nothing intermediate to these two things."

This is a serious difficulty. The critical point lies in the question: can it be *proved* that a nest of intervals contracts to a point? Or is its existence merely a hypothesis? In the meantime we will take up another standpoint regarding this matter. Namely, we will learn to regard these *nests of intervals themselves as numbers*. However, before this end can be reached we must occupy ourselves with a somewhat more modest question.

We have stated that a sequence $a_1, a_2, a_3 \ldots a_n, \ldots$ is convergent if there is a number a which the terms of the sequence approach without restriction. Though this definition is actually in order, it has a drawback, since it assumes that we *know* the limiting value. Now, it is often much more difficult to find the limiting value than to test the behavior of the sequence. If the reader looks at the graph of sequence 8 on p.133, he will immediately have the impression of convergence, although he may not be able to compute the limiting value. It is therefore desirable to free the concept of convergence from that of limiting value. Accordingly we ask: can the concept of convergence be so formulated that it can be decided on the basis of the behavior of the sequence of numbers alone whether it converges?

This is actually possible. Let us assume that the sequence of numbers $a_1, a_2, a_3, \ldots a_n, \ldots$ converges to the (rational) limiting value a. Then an index N can be found such that

$$|a - a_n| < \frac{\varepsilon}{2}, \quad \text{if} \quad n > N.$$

For every term a_{n+p} lying further away it is valid all the more that

$$|a - a_{n+p}| < \frac{\varepsilon}{2}, \quad \text{if} \quad n > N.$$

From these two relations a new one can be derived in which

the number a will not appear. Thus, let us consider the difference $a_n - a_{n+p}$; the value of this difference will not change if we add and then subtract a from it; therefore,

$$a_n - a_{n+p} = a_n - a + a - a_{n+p} = (a - a_{n+p}) - (a - a_n).$$

If we now form the absolute value of this expression and take into account the remark on p.142, then

$$\begin{aligned}|a_n - a_{n+p}| &= |(a - a_{n+p}) - (a - a_n)| \\ &\leq |a - a_{n+p}| + |a - a_n| \\ &< \frac{\varepsilon}{2} + \frac{\varepsilon}{2} = \varepsilon.\end{aligned}$$

Sequences which behave in this way will be called *convergent*. Consequently, the definition is:

The sequence $a_1, a_2, \ldots a_n, \ldots$ is called convergent if

$$|a_n - a_{n+p}| < \varepsilon, \text{ provided that } n > N.$$

This is an *immanent* criterion, that is, a criterion which refers only to the terms of the given sequence and makes the process of stepping over the stock of these terms unnecessary. Intuitively speaking, the criterion says that the terms of the sequence must be so clustered that the difference between one term and *any* other later one turns out to be arbitrarily small provided that the term lies sufficiently far out; or, still simpler, that all terms which are far out lie very close together.

As the reader knows, there are sequences which are convergent but tend toward an irrational limiting value, for example, the sequence that converges to $\sqrt{2}$. However, since up to now we have at our command only the rational numbers, we are strictly speaking, not allowed to act as if there were irrational limiting values. To be precise, we have to say instead that there are two kinds of convergent sequences: first, those which have rational limiting values, secondly, those which do not have rational limiting values. This is only another way of expressing the fact already known to us that the rational number system is not closed relative to the limit operation.

13. THE REAL NUMBERS

Accordingly, we are again confronted with the problem of extending the domain of numbers. How this is to be done has been clearly pointed out to us by the previous considerations, namely, we will construct a calculus with sequences in accordance with the model of Chapter 11. The elements of our consideration will therefore be sequences of rational numbers, more strictly, convergent sequences in the sense explained above. We will show that these sequences can be ordered by means of the relations "greater," "equal," "smaller," and that calculating operations can be unambiguously defined for these sequences. This will finally give us the right to designate the sequences themselves as a new kind of number, as "real numbers."

When introduced in this way the difficulties which burdened the old interpretations of the concept of irrational number vanish all at once. There are nowadays several theories of irrational numbers. In the following we will sketch the construction, A according to Cantor, B according to Dedekind.

A. Cantor's Theory

DEF. 1. A sequence $a_1, a_2, a_3, \ldots a_n, \ldots$ is called a *null sequence* if for every ε there exists an N such that

$$|a_n| < \varepsilon \text{ provided that } n > N.$$

Consequently the sequence $0, 0, \ldots 0, \ldots$ is a null sequence.

THEOREM 1. If (a_n) is a null sequence, then $(-a_n)$ is also a null sequence.

THEOREM 2. If (a_n) and (b_n) are null sequences, then $(a_n + b_n)$ is also a null sequence.

DEF. 2. Two sequences are said to be *equal* if their difference sequence is a null sequence. In symbols: $(a_1, a_2, \ldots a_n, \ldots) = (b_1, b_2, \ldots b_n, \ldots)$, if $(a_1 - b_1, a_2 - b_2, \ldots a_n - b_n, \ldots)$ a null sequence.

This definition satisfies the requirements which we placed on the concept of equality; it is

a) reflexive: $(a_n) = (a_n)$, for $(a_n - a_n) = (0)$ is a null sequence;

b) symmetric: from $(a_n) = (b_n)$ follows $(b_n) = (a_n)$, for $(a_n - b_n)$ and $(b_n - a_n)$ are null sequences at the same time;

c) transitive: from $(a_n) = (b_n)$ and $(b_n) = (c_n)$ follows $(a_n) = (c_n)$; for if $(a_n - b_n)$ and $(b_n - c_n)$ are null sequences, then the sequence $(a_n - c_n)$ resulting from addition is also a null sequence.

This definition implies that there are always infinitely many sequences of numbers which are equal to a given sequence. (This reminds us of the fact that there are infinitely many number couples which represent the same rational number.)

We now come to the concepts "positive" and "negative."

DEF. 3. A sequence $a_1, a_2, \ldots a_n, \ldots$ is said to be *positive* if there is a positive rational number r such that almost all terms of the sequence lie to the right of r. A sequence is said to be *negative* if there is a negative rational number s such that almost all terms of the sequence lie to the left of s.

It must now be proved that the properties of being positive, negative and null sequences, are incompatible and that they form an exhaustive disjunction. We express this in the following two theorems.

THEOREM 3. A sequence cannot be a positive as well as a negative or a null sequence.

Namely, if a sequence is positive, then almost all its terms lie to the right of a fixed positive number r; but in such a case they cannot become arbitrarily small; therefore, the sequence certainly cannot represent a null sequence and still less a negative sequence. Likewise we recognize that if a sequence is negative, it is not a null sequence, and much less a positive one. Finally, since if a sequence is a null sequence its term must become arbitrarily small, it follows that neither a positive nor a negative constant can be found such that almost all its terms lie to the right of the one or to the left of the other.

THEOREM 4. *These three possibilities form an exhaustive disjunction, that is, a convergent sequence is either a positive or a negative or a null sequence.*

For either there is a positive number r such that almost all terms a_n lie to the right of r; or there is a negative number s such that almost all terms lie to the left of s; or neither case arises. The reason, for example, could be that there are infinitely many terms of the sequence to the right of r and infinitely many others to the left of s, which means that the terms leap to and fro; however, this would be incompatible with the convergence of the sequence. Hence there remains only the possibility that the terms of the sequence lie neither almost all to the right of r nor almost all to the left of s, but that they lie between r and s. Now, since r and s are to be entirely arbitrary numbers, this means that almost all terms are situated in any arbitrary (therefore also any arbitrarily small) interval around 0, and consequently form a null sequence.

However, another question suddenly turns up. According to a remark made earlier there are always infinitely many sequences which are equal to one another. Now does the positiveness of a sequence carry over to the sequences which are equal to it? In other words, is positiveness a characteristic which adheres, let us say, to the particular form of the sequence and is lost if the sequence is replaced by any other which is equal to it? The answer is given by

THEOREM 5. *If (a_n) is positive and $(a_n) = (b_n)$, then (b_n) is also positive.*

Namely, if (a_n) is positive, then almost all terms lie to the right of a certain positive number r; but this must also be true of the terms of the sequence (b_n), since these approach closer and closer to the terms of the first sequence.

What we have just proved regarding the positive sequences is valid in exactly the same way if the sequence is a negative or a null sequence.

For the following we still need two further theorems.

THEOREM 6. If the sequence (a_n) is positive, then the sequence $(-a_n)$ is negative.

THEOREM 7. If (a_n) and (b_n) are both positive sequences, then $(a_n + b_n)$ is also positive.

We now proceed to the definition of "greater" and "smaller."

DEF. 4. The sequence (a_n) is said to be *greater* than the sequence (b_n), in symbols

$$(a_n) > (b_n)$$

if the difference sequence $(a_n - b_n)$ is positive.

DEF. 5. $(a_n) < (b_n)$ if $(a_n - b_n)$ is negative.

The relations so defined are a) irreflexive, b) asymmetric, c) transitive.

As to a). No sequence is greater than itself; for, if we had $(a_n) > (a_n)$, this would mean that $(a_n - a_n)$, which is the sequence (0), would be positive, whereas it is a null sequence.

As to b). If $(a_n) > (b_n)$, then $(a_n - b_n)$ is positive (Def. 4); but then $(b_n - a_n)$ is negative (Theorem 6) and consequently $(b_n) < (a_n)$ (Def. 5).

As to c). If $(a_n) > (b_n)$ and $(b_n) > (c_n)$, then $(a_n - b_n)$ and $(b_n - c_n)$ are both positive; by Theorem 7, the sequence $(a_n - b_n + b_n - c_n) = (a_n - c_n)$ arising by addition is also positive, which means: $(a_n) > (c_n)$.

We are now far enough to be able to prove the following basic theorem.

THEOREM 8. The convergent sequence form an ordered system.

Namely, if two arbitrary sequences (a_n) and (b_n) are selected, then (a_n) shall either be greater, equal or smaller than (b_n); the case of incomparability, which perhaps coul be thinkable, shall therefore be excluded. This simply goes back to the fact that $(a_n - b_n)$ is either a positive, negative or a null sequence, and that these three possibilities form an exhaustive disjunction.

After this orientation regarding the fact that convergent sequences can be ordered, we will now take up the definition of the calculating operations.

Definition of sum: $(a_n) + (b_n) = (a_n + b_n)$.

The reader will remember the conditions we imposed on the concept of sum.

a) The sum shall exist; which means that the sum of two convergent sequences shall again be a convergent sequence.

b) The sum shall be uniquely determined; if we replace (a_n) and (b_n) by any other sequences (a_n') and (b_n') equal to them, then $(a_n + b_n) = (a_n' + b_n')$

c) It shall obey the associative and

d) the commutative laws.

The proof that these conditions are satisfied is omitted here; some of the details are indicated in the chapter on operating with sequences. We will now go on with the definitions of the other calculating operations:

$$(a_n) - (b_n) = (a_n - b_n)$$
$$(a_n) \cdot (b_n) = (a_n \cdot b_n)$$
$$(a_n) : (b_n) = \left(\frac{a_n}{b_n}\right)$$

It can be shown that these calculating operations possess the same formal properties as the like-named operations in the domain of rational numbers.

Consequently, we conclude that we can compare sequences and operate with them just as we did with rational numbers.

Hence there is no longer anything which prevents us from designating the convergent sequences as a new kind of number, as *real* numbers. Now, how are the rational numbers related to the real numbers? Again we must avoid the error of inserting

13. THE REAL NUMBERS

the rational numbers as a part of the reals. What we can do, however, is to set up a correspondence between the rational numbers and a *subclass* of sequences. Namely, to every rational number r we can associate the sequence

$$(r, r, \ldots r, \ldots)$$

—or any other sequence which converges to r; and we recognize at once that two such sequences, for instance $(r, r, \ldots, r, \ldots)$ and $(s, s, \ldots, s, \ldots)$, are related to one another in exactly the same way as the rational numbers r and s. Thus, the two sequences are said to be equal if r and s are equal; the first sequence is said to be greater than the second, if r is greater than s; the sum of the two sequences corresponds to the sum of the numbers r and s. In short, the rational numbers and these particular sequences (the sequences with rational limiting values) are associated to one another by a one-to-one, similar and isomorphic correspondence. This is evidently the reason why the rational numbers seem to be a special case of the reals. However, we will clearly stress the distinction by distinguishing between the rational number r and the rational *real* number $(r, r, \ldots, r, \ldots)$.

On designating the sequences as numbers, we are merely adhering to a principle which was after all our guide throughout the whole time, directing us in the formation of the concepts of integer and rational number. Nevertheless, the idea of regarding the sequences themselves as numbers, when it emerged first, contained something so bold that many mathematicians shrank back before it. A sequence, so they perhaps said, is, after all, something that is composed of infinitely many numbers; it has not the slightest similarity to the concept of a quantity which can be represented on the number axis. And, if the sequence is interpreted as a rule or as a law for the generation of numbers, it belongs more than ever to a logical category other than that of the numbers themselves—is it not absurd to calculate with rules or laws?

13. THE REAL NUMBERS

This struggle to ward off the "formal theory" of irrational numbers was very drastically waged by du Bois-Reymond. "A purely formalistic-literal framework of analysis," so we read in the writings of this author, "which is what the separation of number from magnitude amounts to, would degrade this science to a mere game of symbols, where arbitrary meanings are attributed to the letters as to chessmen and playing cards. Amusing as such a game could be, this literal mathematics would soon enough exhaust itself in fruitless tendencies. No doubt, with help from so-called axioms, from conventions, from philosophemes contrived *ad hoc*, from unintelligible extensions of originally clear concepts, a system of arithmetic can be constructed which resembles in every way the one obtained from the concept of magnitude, in order thus to isolate the computational mathematics, as it were, by a cordon of dogmas and defensive definitions from the psychological domain. Also, unusual ingenuity may have been spent on such constructions. However, other arithmetical systems could also be invented in the same way. Ordinary arithmetic is just the one which corresponds to the concept of linear magnitude"[2] (here the author means the concept of measurable quantity). It is also very remarkable that H. Hankel, the creator of a *purely formal* theory of rational number, turned sharply against these theories: "Every attempt to treat the irrational numbers formally and without the concept of magnitude must lead to the most abstruse and troublesome artificialities, which, even if they can be carried through with complete rigor, as we have every right to doubt, do not have a higher scientific value."[3]

To these authors, therefore, a sequence does not seem to agree with what they understand by a real number. They are

[2] *Die allgemeine Funktionentheorie* (*General Theory of Functions*), p. 53f.

[3] *Theorie der komplexen Zahlensysteme* (*Theory of the Systems of Complex Numbers*), p. 46.

13. THE REAL NUMBERS

rather inclined to *distinguish* the sequence, as the approximation process, from the limit, the irrational number, and then, to be sure, stand before the difficulty of explaining why the introduction of irrational numbers is justified.

However, basically it is only the obscurity of thinking, psychological difficulties, which make this matter look like a problem. What do we actually mean when we say that we *know* an irrational number, for example $\sqrt{2}$, and that we have a conception of its magnitude? What is at the bottom of this feeling? Certainly no more than the knowledge of a process for computing $\sqrt{2}$ to arbitrarily many decimals. To know an irrational number means to know a process for computing it approximately. In this sense it is completely proper to identify the irrational numbers with the approximation process (the sequence). If, as an example of a logical problem, the question has been raised: "Does a sequence merely *approach* its limiting value or actually *reach* it?" we can now very well leave the answer to the reader himself.

A second reason which detained many from this equalization is that an irrational number is visualized as a point on the number axis. From this standpoint it is found difficult to say that a number is a law or a rule. What leads us astray here is the use of too crude and primitive a simile. We visualize, perhaps, that a real number is obtained by reaching into the set of real numbers and drawing one out; we give no consideration whatever to the fact that the irrational number is given only by a *construction*, for instance, by a convergent process from which, as matters stand, it cannot be severed. Actually the nature of an irrational number can best be rendered by saying that it is a rule for the generation of rational numbers.

We have now to consider a modification of this interpretation which comes somewhat closer to the idea we usually have of an irrational number. Hitherto we have permitted arbitrary (convergent) sequences without placing much importance to whether the sequences were increasing or decreasing or oscillating. Now we wish to confine our considerations to "monotone" sequences, that is, to sequences whose terms vary only in *one* sense. We think of two such monotone sequences

$$a_1\ a_2,\ \ldots\ a_n,\ \ldots$$
$$b_1\ b_2,\ \ldots\ b_n,\ \ldots$$

as formed with the following properties:

1. The first sequence is monotonic increasing, which means that

$$a_1 \leq a_2 \leq a_3 \leq \ldots$$

2. The second sequence is monotonic decreasing, which means that

$$b_1 \geq b_2 \geq b_3 \geq \ldots$$

3. No term of the first sequence is greater than the corresponding term of the second sequence, therefore,

$$b_1 \geq a_1,\ b_2 \geq a_2,\ b_3 \geq a_3,\ \ldots$$

4. The differences $b_n - a_n$ shall become arbitrarily small with increasing n.

We now wish to combine a pair of such sequences and designate the pair by a symbol, say $\left(\dfrac{a_n}{b_n}\right)$. It is not difficult to transfer to such pairs of sequences the definitions and operating rules which were previously defined for sequences; instead of using convergent sequences we could erect a calculus with pairs of sequences. Geometrically, such a pair of sequences means a nest of intervals. The points $a_1, a_2, a_3, \ldots$ move further and further to the right

he points $b_1, b_2, b_3, \ldots$, further and further to the left; consequently they define a series of intervals which draw together without restrictions. Hence it means only a slight variation of Cantor's theory to have *such a nest of intervals stand for a real number*.

In many investigations the real number must be formulated in just this way. We cite only two examples.

1. An infinite decimal fraction is no more than an abbreviated expression for such a nest of intervals. If we find, for example, in the computation of $\sqrt{2}$

$$\sqrt{2} = 1\cdot 41421 \ldots,$$

then we can also write this in the form

$$\sqrt{2} = \begin{pmatrix} 2, & 1\cdot 5, & 1\cdot 42, & 1\cdot 415, & \ldots \\ 1, & 1\cdot 4, & 1\cdot 41, & 1\cdot 414, & \ldots \end{pmatrix}$$

We thereby see that through the decimal fraction development of $\sqrt{2}$ a law is set up by which intervals (the consequent of each interval is a tenth of the antecedent) are nested within one another.

2. If we wish to compute the area of a circle by means of elementary geometry, the following procedure, as is well known, can be used. On the one hand, let a regular hexagon, dodecagon, icositetragon . . . be successively inscribed in the circle; on the other hand, let a regular hexagon, dodecagon, icositetragon . . . be circumscribed about the circle. Thereby the area of the circle is squeezed into an interval which becomes narrower and narrower. The area of the inscribed polygons forms a monotonic increasing sequence; the area of the circumscribed polygons, a monotonic decreasing sequence. Furthermore, no term of the first sequence is greater than a term of the second and the difference between the area of a circumscribed and the corresponding inscribed polygon becomes less than any positive value if the process is continued. Hence our four conditions are satisfied and it follows that the *area of the circle is represented by a nest of intervals*, as discussed here. (Incidentally, one should not take this as a proof that the are

of a circle has such and such a magnitude; rather, the area is *defined* by saying that it is the common limit approached by the inscribed and circumscribed polygonal areas. The concept of area has been explained up to now only for figures with straight-line boundaries and must be defined anew for curvilinear boundaries.)

The process of nesting intervals is often very effective to prove the existence of certain numbers. The proof of the theorem of Bolzano-Weierstrass may serve as an example:

Every bounded infinite point set has at least one point of accumulation.

The adjective "infinite" says that the set consists of infinitely many points; the adjective "bounded," that all these points lie between two fixed bounds a and b such that the whole set occupies a finite portion of the straight line.

The proof consists in bisecting the interval (a, b) and then selecting that half which contains infinitely many points. (In case infinitely many points lie in each subinterval, we choose one arbitrarily.) With the subinterval so chosen we continue in exactly the same way, namely, we again divide it in half and mark that subinterval which contains infinitely many points. Continuing in this way we obtain a series of intervals nested within one another whose lengths decrease without restriction and which contract to a point. This point is a point of accumulation, for its construction is certainly such that each of the subintervals in which it lies contains infinitely many points of the set. If it were to happen after the division of some interval that infinitely many points of the set exist in each of the two halves, then the process branches out. In such a case we can continue the process with either half of the interval and recognize that there are then different points of accumulation.

Another question is whether we can always *decide* which subinterval contains infinitely many points. We assume that

13. THE REAL NUMBERS

a line segment contains either a finite or an infinite number of points of the set, that this is objectively established, no matter whether we can determine it or not. In recent years this inference has been subjected to criticism; the intuitionist, for instance, attacks the legitimacy of such a procedure. For him a statement has a sense only if it can be checked in finitely many steps. Since the theorem of Bolzano-Weierstrass is the supporting pillar of the whole of analysis, we can imagine how far-reaching this criticism is. The pursuit of these questions, however, would lead us beyond the scope of this book.

We had set ourselves the task of filling up the gaps among the rational numbers, that is, of making the limit operation performable without restrictions. This purpose is attained, since every convergent sequence of rational numbers $a_1, a_2, \ldots a_n, \ldots$ has a limiting value, the real number a, where a is defined by the convergent sequence itself. However, this poses a further question. What happens if we repeat this process in the domain of real numbers? that is, if we form sequences whose terms are real numbers? Will we thereby perhaps be led to new numbers which are unattainable by the foregoing methods? This question is of considerable importance for mathematics. If it were to be answered in the affirmative, the structure of the real number domain would then be as follows: first, a layer of real numbers of first degree, defined as convergent sequences of rational numbers, would be superimposed on the system of rational numbers; upon these numbers of the first degree numbers of second degree would be erected, the convergent sequences of real numbers of the first degree, etc. The totality of the real numbers would accordingly form a hierarchy, and we could no longer speak of real numbers plain and simple. It is evident that the structure of arithmetic would thereby become much more involved. Thus, in any form of analysis we would first have to test whether it was valid for numbers of this or that degree, and it would perhaps be impossible to pronounce general laws. Fortunately, this distinction is not necessary. For, the proposition is valid that every convergent sequence of real numbers can be replaced by a convergent sequence of rational numbers; therefore, the formation of convergent sequences in the domain of real numbers leads to nothing new.

Before proceeding, I will make a remark. A sequence of real numbers is a new symbol whose use has not yet been explained. We now stipulate that the rules which were set up for convergent sequences of rational numbers shall remain in effect for convergent sequences of real numbers, too; in particular, therefore, two such sequences shall be called *equal* if their difference sequence is a null sequence.

Now, in order to see the truth of the statement, let us form an arbitrary convergent sequence of real, for example, irrational numbers

(I) $$a_1, a_2, a_3, \ldots a_n, \ldots$$

Each of these numbers is defined as a sequence of rational numbers,[4]

$$a_1 = (a_{11}, a_{12}, \ldots a_{1n}, \ldots)$$
$$a_2 = (a_{21}, a_{22}, \ldots a_{2n}, \ldots)$$
$$\vdots$$
$$a_m = (a_{m1}, a_{m2}, \ldots a_{mn}, \ldots),$$

so that the sequence (I), strictly speaking, stands for a sequence of sequences of rational numbers. The statement now amounts to saying that this sequence of sequences can be replaced by a simple sequence. Hence we must be able to give a sequence of rational numbers whose terms differ as little as we wish from the terms of the given irrational sequence; then one sequence can be replaced by the other. We now determine an index N_1 in the first sequence such that from there on the

[4] In the following table we use double indices.

13. THE REAL NUMBERS

difference of the terms of the sequence from α_1 comes out less than 1;[5] in the second sequence we go up to an index N_2, from whence on the difference from α_2 is less than $1/2$, etc. From each of these sequences let us sort out the term with the appropriate index and for a sequence of *rational* numbers.

(II) $\qquad\qquad a_{1N_1}, a_{2N_2}, a_{3N_3}, \ldots$

If we now compare the sequences (I) and (II) with each other, their differences

$$\alpha_1 - a_{1N_1}, \alpha_2 - a_{2N_2}, \alpha_3 - a_{3N_3}, \ldots$$

form a sequence whose terms are successively smaller than

$$1, 1/2, 1/3, \ldots;$$

consequently, it is a null sequence. Accordingly, the sequences (I) and (II) are equal in the sense of our definition.

This result can also be expressed as follows: *the system of real numbers is closed relative to the limit operation.* Thereby it has been made clear that this extension is formally similar to the two earlier extensions of the number domain.

B. Dedekind's Theory

A theory of another style is due to Dedekind. Dedekind was led to these considerations by the fact that he could never give a clear definition of continuity in his lectures. It seems intuitively quite clear what is meant in saying that a straight line is continuous; but to the mathematician this is not enough, since he seeks a precise criterion to give his deductions a point of application. Naturally, nothing is gained by vague remarks about points adhering to one another, about lines being gapless down to their smallest parts, etc., since such statements cannot serve as the basis for valid conclusions. Hence Dedekind r

---

[5] i.e., such that the difference $(a_{11}, a_{12}, a_{13}, \ldots) - (a_{1N_1}, a_{1N_1'}, a_{1N_1''}, \ldots) <$

solved to meditate on the question of continuity until finding a conceptually clear definition. Finally, he found what he sought. His train of thought is as simple as it is original. Let the reader visualize a straight line (say, a horizontal one) and choose a point on it at will. This point then separates the straight line into two portions such that any point of one portion lies to the left of any point of the other. As to the point itself, it can be included, according to our preference, in the one or the other portion. Dedekind now finds the essence of continuity in the converse, namely, in the following principle:

"If all points of the straight line fall into two classes such that every point of the first class lies to the left of every point of the second class, then there exists one and only one point which produces this division of all points into two classes, this severing of the straight line into two portions."

Dedekind continues: "As already said I think I shall not err in assuming that everyone will immediately grant the truth of this statement; the majority of my readers will be very disappointed in learning that by this commonplace remark the secret of continuity is to be revealed. To this I may say that I am very glad if everyone finds the above principle so obvious and so in harmony with his ideas of a line; for I am utterly unable to adduce any proof of its correctness, nor has anyone the power. The assumption of this property of the line is nothing else but an axiom by which only we attribute to the line its continuity, by which our thinking endows the line with continuity."

The meaning of this concept will immediately stand out if we ask ourselves the question: is the system of rational numbers continuous in the sense explained above? The rational numbers are dense, as is well known, and the opinion may easily arise that dense means the same as continuous. The criterion of Dedekind shows at once that this is not so. For, if we assume a separation of the rational numbers into two

13. THE REAL NUMBERS

classes according to the following principle: in the class to the left we assign all rational numbers which are negative, also zero and all positive numbers whose square is smaller than 2; in the class to the right we assign all other numbers, then we have effected a complete separation of the rational number system, by which every rational number falls in one and only one class. But is this separation produced by a rational number? That is, is there a rational number which is either the greatest element of the lower class or the smallest element of the upper class? No! For, at the exact point where the two classes meet, so to speak, there can be no rational number, since $\sqrt{2}$ is irrational. Hence there are separations of the rational numbers into two classes which are not produced by a rational number. (Naturally separations also exist which are produced by a rational number. For example, if we assign to the lower class all rational numbers to the left of 1 and to the upper class all rational numbers from 1 up, then the two classes will be exactly separated by the number 1.) At any rate, there are separations which do not correspond to a rational number; therefore, this would reveal the discontinuity of the rational number system even though we had never heard of irrational numbers.

On the other hand, in the real numbers we encounter an example of a continuous system. Before going into this, however, we must introduce some concepts. Let us think of a system of entities (numbers, points ...) which are completely arbitrary (thus, they need not be quantitative) except that we assume that they are ordered. Then, for any two entities it is well established which is the antecedent and which is the consequent. We now consider separations of the entities of the system into two classes A and B which have the following properties:

1. No class shall be empty (this means that it shall never happen that all entities lie only in one class).

2. Every entity of the class A shall precede in the ordering the entities of class B.

13. THE REAL NUMBERS

3. Each entity shall belong to one and only one class.

Such a separation is called a "cut" and is designated by (A/B). In terms of the boundaries of the two classes four cases are possible.

1. A has a last, B has a first element.
2. A has a last, B has no first element.
3. A has no last, B has a first element.
4. A has no last, B has no first element.

In the first case the cut is called a *jump*; in the second and third cases, *continuous*; in the fourth case, a *gap*. Accordingly, we can formulate Dedekind's definition as follows: an ordered system is continuous if every cut is continuous, that is, if neither jumps nor gaps appear. The reader should note once more that the application of this definition is not restricted to numbers. He may also think of it in terms of brightness, musical pitch, etc. However, it is well for him to render this concept clear first by means of number examples.

1. The series of integers exhibits jumps but no gaps. For, if we separate these numbers somehow into two categories—for example, all numbers which are smaller than 2 by A, all numbers greater than 2 by B—then the lower class will have a greatest number and the upper class a smallest.

2. The system of rational numbers exhibits gaps but no jumps (cf. the example mentioned on the previous page).

The latter constitutes what is meant when we say that the system of rational numbers possesses gaps or is discontinuous. Its stock of numbers is just too scanty to provide that every cut could be generated by one of its numbers. In the system of rational numbers there are more cuts than numbers. This brings us now to the idea of extending the number system so that the cuts themselves are interpreted as numbers in a new sense. Just as in our previous proceedings, we seek to construct a calculus with cuts. In the following we will sketch its first steps.

13. THE REAL NUMBERS

DEF. 1. Two cuts (A/B) and (A'/B') are said to be equal if A coincides with A' and B with B'.

If we separate the system of rational numbers at the position 2, then we can assign the number 2 at pleasure to the lower or the upper class. *Two such cuts are regarded as not essentially different*; in the sense of our definition, however, they would be unequal. For this reason we will improve the definition by stating that two cuts are said to be equal if their classes coincide with the exception of at most one number.

DEF. 2. The cut (A/B) is said to be smaller than the cut (A'/B') if there are rational numbers which belong to A' but not to A.

The definition of "greater" is entirely analogous.

DEF. 3. A cut is said to be positive if its lower class contains at least one positive number.

DEF. 4. A cut is said to be negative if its upper class contains at least one negative number.

DEF. 5. A cut is said to be a null cut if there is no positive number in the lower class and no negative number in the upper class.

DEF. 6. The sum of two cuts (A/B) and (A'/B') is the cut whose lower class consists of all sums that can be formed by adding a number in A to a number in A', and whose upper class is similarly obtained from the numbers in B and B'.

These few strokes may be sufficient to give the reader a picture of the construction of this theory. Naturally he must realize that in the exact construction it must be *proved* that the concepts of equality, less than, sum, etc. possess the formal properties which these concepts have in the arithmetic of rational numbers.

Next, by a rational real number we will understand a cut which is produced by a rational number, *not* the rational number itself. Between the rational numbers and the associated cuts there only exists a correspondence which is one-to-one similar and isomorphic.

Now what happens if we undertake new separations in the domain of the real numbers (the cuts) thus obtained? Would we thereby leave the domain? Well, it is fundamentally important that this is not the case. Instead the following proposition is valid: every cut in the domain of real numbers can be replaced by a cut in the domain of rational numbers. In order to be able to compare two such cuts, we must first explain when they are to be called equal. If we designate a cut in the system of real numbers by ($\mathfrak{A}/\mathfrak{B}$), a cut in the domain of rational numbers by (A/B), then $\mathfrak{A}$ will consist only of real numbers, A only of rationals. For comparison we restrict ourselves to the *rational* real numbers occurring in $\mathfrak{A}$. We now define: the two cuts are equal if their classes coincide in so far as the rational numbers are concerned.

Now, from a real cut ($\mathfrak{A}/\mathfrak{B}$) a rational cut (A/B) can be obtained by the following rule: a rational number r is contained in the lower class if the corresponding rational real number occurs in the lower class $\mathfrak{A}$; it will be contained in B, if the corresponding number occurs in the upper class $\mathfrak{B}$. The cut (A/B) so formed is obviously equal to the cut ($\mathfrak{A}/\mathfrak{B}$).

Hence we obtain the result: every cut in the domain of real numbers is itself a real number. The cut formation does not lead us out of the domain and this means that *the real numbers form a continuous system, the real continuum.*

We again wish to base an example on the construction of the abstract theory in order to illustrate how one actually works with these concepts. As is well known, an infinite point set need have neither a maximum nor a minimum, even though it occupies a bounded portion of the number axis—we only have to think of the proper fractions between 0 and 1. However, such a set always has a least upper and a greatest lower bound; in our case the numbers 0 and 1. These are defined as follows. Let us call a number a minorant if it is surpassed by all elements of the set; graphically speaking, this means a

203

point which lies to the left of the entire point set. Evidently, there are infinitely many such minorants. If we think of taking the greatest of all minorants, then we obtain the greatest lower bound. This is, therefore, a number which is surpassed by all terms of the set but which loses this property as soon as it is replaced by a somewhat greater number. We are now confronted with the following mathematical problem: does the greatest lower bound *exist?* Does the set of minorants have a maximum? In order to prove this we divide the rational numbers into two classes. In the lower class we assign all numbers which lie to the left of the totality of numbers of the set; in the upper class, only those which will surpass at least one number of the set. It is clear that this division has the properties of a cut and therefore defines a real number g. This number g cannot be greater than any number of the set. For, if it were greater than a number of the set, let us say a, then there would be rational numbers in the lower class of g that lie to the right of a, whereas this class was defined so that it should contain no rational number of this sort. Accordingly, all numbers of the set lie to the right of g; consequently, g is a minorant. If it is replaced, however, by a somewhat greater number $g + \varepsilon$, then is ceases to be a minorant. For, in the lower class of $g + \varepsilon$ there would now also be numbers which earlier were located in the upper class of g; but these are numbers which surpass at least one element of the set; consequently $g + \varepsilon$ is no longer a minorant. g is accordingly a minorant such that there is no greater one; this means that it is the maximum of the minorants. — If the reader should find it difficult to follow this proof, then there is a good method for remedying this. Let him try to prove the existence of an upper limit in an entirely corresponding manner. He will thereby be forced to produce the whole chain of ideas once more, and then he will understand what it is all about.

Dedekind's theory may be modified. Since the upper class of a cut is uniquely determined by the lower class, then the latter suffices for the determination of a real number. We will call it a *segment*. It is not difficult to transfer all definitions to segments. Russell constructed the real numbers in this way.

C. Comparison of the Two Theories

Let us now compare the systems of Cantor and Dedekind. The structures of both theories were called real numbers. Now, strictly speaking, is this justifiable? Are they the same numbers? Here it is certainly not a question of an identity, since the structures are defined quite differently. It is true, however, that there exists an exceptionally close correspondence between them, since the number systems created by Cantor and Dedekind can be related to one another by a *one-to-one, similar and isomorphic* correspondence. Hence they represent number systems which have the very same structure. We will recognize this at once by means of the following simple consideration. Cantor's theory operates with sequences or, as we would rather now say, with nests of intervals. We have seen that there are infinitely many distinct nests of intervals that contract to one and the same point. If we think of all these nests as superimposed one over the other (say, on the number axis), then all left endpoints form the lower class, all right endpoints the upper class of the Dedekind cut. Hence the cut is nothing but a kind of condensation of all possible nests of intervals which shrink up to a point.

From Dedekind's standpoint we can prove that a nest of intervals determines a point; from Cantor's standpoint we can prove that to every cut there corresponds a real number. Both theories are independent of each other and each theorem of analysis can be formulated just as well in one as in the other theory. This is the reason why in mathematics we speak simply of real numbers and place no importance on whether they are numbers in the sense of Cantor or of Dedekind.

13. THE REAL NUMBERS

The basic element of Cantor's theory is the sequence, the basic element of Dedekind's theory is the class. This imparts a certain advantage to Dedekind's theory for it determines a real number by a single cut, whereas in Cantor's theory there are various ways of representing a number. In the case of the conventions set up for the concepts "greater," "equal," "smaller," "positive," "sum," etc., we must prove in Cantor's theory that the formal properties of the relations and operations are independent of the particular choice of sequence; in Dedekind's theory, however, this necessity is eliminated. Hence Dedekind's theory gets along with fewer tools, it goes to work more economically and from this point of view appears simpler.

From another standpoint, however, we may prefer Cantor's theory, for Dedekind can handle only *linear* continua, while Cantor can handle continua of arbitrary dimension (nests of intervals in the plane, in space, etc.).

Dedekind's theory led to a surprisingly simple characterization of the continuity of a straight line. Let us finally ask how would the same fact have to be formulated from the starting point of Cantor?

A point set could be thought of as continuous if it is closed under the limit operation, that is, if it contains each of its points of accumulation. That this is not true is shown by the set

$$0, 1, \frac{1}{2}, \frac{1}{3}, \frac{1}{4}, \ldots \frac{1}{n},$$

which is closed but not continuous. This evidently rests on the fact that it consists, except for 0, *only* of isolated points (that is, points which have a neighborhood free of points of the set. Hence, for a set to be continuous we will require that every point of the set be surrounded by neighboring points in arbitrarily close neighborhoods or, in other words, that every point of the set be a point of accumulation. The nature of the continuous connection of a set seems, therefore, to be contained in the following two properties:

1. Every point of the set is a point of accumulation.
2. Every point of accumulation of the set belongs to it.

In the theory of sets the totality of points of accumulation of a set M is called its derived set and is designated by M^1. The two properties can then be expressed as follows:

1. The set is wholly contained in its derived set, namely, M is a part of M^1.
2. The derived set is wholly contained in the set, namely, M^1 is a part of M.

This amounts to saying that M and M^1 coincide. Sets of this kind are called *perfect*. Cantor first believed that with the property of "perfect" he had hit upon what we commonly call "continuous." However, he discovered later that there *are perfect sets which are nowhere dense*. This means that they have such a loose texture that no interval, no matter how small, on the straight line can be exhibited, in which the points lie densely.

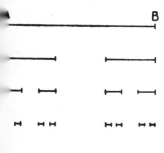

Figure 24

In order to set up such a set we start with an arbitrary line segment AB and delete its middle third with the exception of the endpoints which are left standing. We repeat this process with each of the two subsegments, namely, we again delete their middle thirds, etc., and continue this process without end. (Cf. the sketch to the left.) In this way a point set is generated which is, so to speak, infinitely finely perforated or porous and has the following properties:

1. It is closed, that is, it contains the totality of its points of accumulation. Since at each step of the construction it is always only the *interior* of an interval which is deleted, whereas the boundary points are left standing, then out of the closed

207

set AB, from which we started, only closed sets can again arise. (This is valid even if the process is continued without end.)

2. Since two deleted intervals can never meet, no isolated points can be generated; consequently, the set consists only of points of accumulation. 1. and 2. together say that the set is perfect.

3. There is no interval with the property that between any two points of the set there always lies another point of the set; for, however small I choose this interval, there will always be pairs of points, namely, the endpoints of the deleted intervals, between which no further point lies. Hence it is nowhere dense.

Therefore, the property "perfect" is not sufficient to characterize the continuity of a straight line. One must require besides that the set be *dense*—only then does the (linear) continuum result.

D. Uniqueness of the Real Number System

The rational numbers were inserted between the integers, the irrationals between the rational numbers, where it should be perfectly clear by now what we mean by "insertion." If the limit operation (or, the equivalent cut formation) is once again introduced in the domain of real numbers, nothing new is obtained. Accordingly, the process of number extension seems to have reached a certain closing point at the stage of the real numbers. For the time being, however, this is only a conjecture. What shall we reply to someone who tells us that it is still conceivable that some day a new operation will be discovered which will force us to a reiterated thickening of the real number system?

Now let us see how the matter stands with regard to this question. First of all, however, we must come to an understanding about the assumptions incident in the statement

13. The Real Numbers

the question, namely, regarding what is to be understood by a "number system." If we ask whether the system of real numbers can be extended by the insertion of new elements, we still have in mind that the extended system should be somewhat similar to that of the real numbers. But what shall this similarity consist in? There are some suggestive conditions which seem to delineate what should be required of a number system (in the extended sense):

1. The system should be ordered.
2. Four operations should be defined in this system which have the same formal properties as the basic rules of arithmetic.
3. The system should contain a proper subset which can be mapped on the rational number system by a one-to-one, similar and isomorphic correspondence (it should therefore represent some sort of extension of the rational number system).
4. Finally, the validity of the Archimedean axiom should be required, namely, if α and β are any two positive numbers of the system and $\alpha < \beta$, then it should be possible to add α often enough so that the sum

$$a + a + \ldots + a$$

eventually surpasses β. Briefly stated, there should always exist a natural number n such that $n\alpha > \beta$.

If we formulate precisely the concept of number system in this manner, the following theorem gives an answer to the question asked at the beginning:

COMPLETENESS THEOREM. *The real numbers form a system which is capable of no further extension if conditions 1 to 4 are to remain valid.*

We designate by R the system of real numbers, by $\overline{R}$ a system which satisfies the four conditions and which contains the totality of the number entities of R. Let us now select an arbitrary number from $\overline{R}$, say $\overline{\alpha}$. By condition 3 the system $\overline{R}$ contains a subsystem which exhibits the same structure as the rational numbers; for the time being, we will call this sub-

209

13. THE REAL NUMBERS

system briefly "the rational numbers of $\overline{R}$." All rational numbers of $\overline{R}$ can now be divided into two classes according to whether they are smaller or greater than $\overline{a}$. The cut so formed defines a real number a which also belongs to $\overline{R}$. But then what could be the difference between a and $\overline{a}$? Both numbers are realized by the same separation into classes and are drawn together so closely that on the number axis they cannot be separated by a finite interval, no matter how small. If at all, they can be distinguished only by actual-infinitesimal quantities. However, the appearance of such actual-infinitesimal quantities contradicts the Archimedean axiom—here we are anticipating a result which will later be demonstrated in detail. Hence every element of $\overline{R}$ is an ordinary real number, so that $\overline{R}$ cannot represent a system more numerous and comprehensive than R.

There is a second question. In order to eliminate the incompleteness of the rational numbers we have arrived at four different constructions. These are

> convergent sequences,
> nests of intervals,
> Dedekind cuts,
> segments.

All these systems have shown themselves to be similar and isomorphic. However, couldn't other paths be taken which would lead us to essentially new number systems? Speaking more precisely: can a system which satisfies conditions 1 to and in which the limit operation (or the cut formation) can be performed without restriction differ in any essential feature from the system of real numbers? This possibility is excluded by the

UNIQUENESS THEOREM. Every such system can be related by a one-to-one, similar and isomorphic correspondence to the system of real numbers. Hence there is essentially only one such system.

13. THE REAL NUMBERS

Let R again designate the system of real numbers, $\mathfrak{R}$ a system of arbitrary entities which satisfies our assumptions. Then $\mathfrak{R}$ must contain a subsystem having the structure of the rational numbers. We now intend to show that to every element of R there can be assigned an element of $\mathfrak{R}$. Thus, let any real number a be selected from R. It can be defined by a cut in the system of rational numbers, and so it can be written as

$$a = (A/B).$$

But an exactly corresponding cut can be made in the rational numbers of $\mathfrak{R}$ (since to every rational element of R there corresponds a rational element of $\mathfrak{R}$). Therefore, we also obtain there a separation

$$(\mathfrak{A}/\mathfrak{B})$$

which, due to the continuity of $\mathfrak{R}$, again represents a number $\mathfrak{a}$ of this system. *Consequently to every real number* $a = (A/B)$ *in R there corresponds an element* $\mathfrak{a} = (\mathfrak{A}/\mathfrak{B})$ *in* $\mathfrak{R}$, and this correspondence is obviously similar and isomorphic. $\mathfrak{R}$ must therefore contain at least as many elements as R (more sharply stated, there must be a subsystem of $\mathfrak{R}$ which can be mapped on R by a one-to-one, similar and isomorphic correspondence), *but at the same time no more*, since it cannot be extended, due to the Completeness Theorem. Consequently, it must essentially coincide with $\mathfrak{R}$. Thereby, however, we recognize that *the system of real numbers created by us is the only possible system which is free of gaps and satisfies conditions 1 to 4.* —

After all this we now ask: does there correspond to the continuum of real numbers something objective in the material external world? Does it have a transient reality, as Cantor would say? We might think that the arithmetical continuum is only the image of the spatial one. Boltzmann and Clifford have pointed out in this regard that perhaps our space, as well as time, is discontinuous, that, for example, all motions in nature would take place in tiny small jumps as in the movies. All

13. THE REAL NUMBERS

observations are also compatible with the assumption of a discontinuous space and a discontinuous time. And the modern ideas of wave mechanics show that it is physically meaningless to speak of geometrical properties of physical space under a certain threshold, namely, that of the order of magnitude of the atom. It was noted by Dedekind that in many respects neither does geometry require a continuous space. If only those points of space are considered as existing which, let us say, can be represented in a definite coordinate system by "algebraic numbers" (cf. p. 16) then this space is discontinuous throughout; "but in spite of the discontinuity and the existence of gaps in this space, all constructions occurring in Euclid's Elements can, so far as I can see, be just as accurately effected as in a perfectly continuous space; hence the discontinuity of this space would not be noticed in Euclid's science, would not be felt at all." Dedekind continues: "However, if anyone should say that we cannot conceive of space actually as anything else but continuous, I should venture to doubt it and call attention to the fact that a far advanced, refined scientific training is demanded in order to perceive clearly the essence of continuity."[6]

As to physical space one has become accustomed to conceding the justification of this concept. All the more obstinately one stands up for the view that intuition is the proper prototype of the mathematical continuum. What shall we think of this view? Can intuition perform what is demanded of it? Let us begin with a very simple question. Are a definite number of colors seen when one looks at a spectrum? Evidently not if the individual color nuances are thought of as colors. Shall we say that infinitely many colors are seen, perhaps as many as there are real numbers? This would be like losing one' way in a false system of concepts. Actually, the color continuum has a structure entirely different from that of the number continuum

[6] *Was sind und was sollen die Zahlen?* (*What are and what purpose serve the numbers?*) p. XII, f.

tinuum. In the case of two real numbers it is uniquely established whether they are equal or different. No matter how close they may lie near one another on the number axis, they are and always will be different numbers. A color, however, runs into another imperceptibly, it blends with it; more accurately stated, it is without meaning to speak of isolated elements out of which the continuum is to be erected. The concept of number is not applicable to formations of this kind, for the first assumption of counting is that the entities to be counted be clearly distinguishable. Language says, therefore, with an entirely proper feeling that one sees *innumerable* colors. Any numerical statement about colors is thereby forbidden, and that is all that can be said.

The situation is similar in the case of the visual space. Let the reader ask himself how many points he observes in his field of vision. If he is inclined to the view that there are infinitely many, then let him think over how this opinion can be justified. Does he possess, let us say, a process whereby a point can be added to any arbitrary set of points, as in the case of the number sequence, 1, 2, 3, 4, etc? For numbers the "and so on" is the *source* from which the number stream flows. But what about the case of points? If he tries to apply the criterion of Dedekind and to map the points of his field of vision on a proper part of themselves, he sees again that this "won't work." Actually it *means* nothing to seek after such a mapping since the only method at the disposal of the mathematician for this purpose, the inductive method, is not applicable here. "I see infinitely many points," means only that it doesn't make sense to say: I see only twenty points, or I see only thirty points, or I see only forty points. This has nothing to do with the infinity of mathematics. The intuitable continuum simply has an essentially different structure. (For this very reason the attempt to visualize graphically, let us say, the set of rational numbers in the space of vision is doomed to failure.)

13. The Real Numbers

After all this we cannot say that the real numbers are preformed for us in nature or in intuition. They are a free creation of our mind—true enough a creation which is *suggested* to the mathematician from various sides. More specifically, the stimulus is derived from two directions: first, from geometry (the discovery of irrationals by Pythagoras); secondly, from numerical computations with decimal numbers, as it gradually became accepted after the Renaissance. The idea of real number automatically arises when all possible decimal fractions are considered; the theories of Cantor and Dedekind merely made this idea clear and precise.

Let us now return once more to the foundations of analytical geometry. The latter rests on the assumption that there exists a relation between the points of a straight line and the real numbers, as a consequence of which to each point there corresponds exactly one number and to each number exactly one point. Can this be proved? No; rather a new axiom goes into effect here, which *requires* that every point of the straight line can be put in a one-to-one correspondence to the real numbers. This axiom of Cantor-Dedekind forms the foundation proper of our ordinary analytical geometry. However, we will meet later a geometry in which this axiom is not valid.

E. Various Remarks

We shall finally point to certain difficulties incident to the concept of real number. The best way to illustrate these difficulties is to use a train of thoughts due to Brouwer. Let us assume that I have formed a sequence of natural numbers $a_1, a_2 \ldots a_n \ldots$ by the following rule: $a_1 = 1$, $a_2 = 2$; for every other index n $a_n = n$, if Fermat's equation $x^n + y^n = z^n$ cannot be solved in terms of integers x, y, z; if, however, there are solutions of this equation for $n > 2$ and ν designates the smallest value of n for which a solution exists, then from the index ν on let all a_n be $=\nu$, that is, $a_\nu = a_{\nu+1} = a_{\nu+2} = \ldots = \nu$. Since Fermat's Problem is not solved, we do not know whether the sequence of numbers $a_1, a_2, a_3 \ldots$ increases without end or whether the growth breaks off after a while so that from then on all further terms are the same. With the help of this sequence of numbers we now form the sequence

$$\left(-\frac{1}{2}\right)^{a_1}, \left(-\frac{1}{2}\right)^{a_2}, \left(-\frac{1}{2}\right)^{a_3}, \ldots$$

which begins with

$$-\frac{1}{2}, +\frac{1}{4}, -\frac{1}{8}, +\ldots$$

If Fermat's equation does not have a solution, this sequence converges to 0; if, however, there is a smallest solution with the exponent an even number, the sequence converges to a positive number; and if this exponent is an odd number, the sequence converges to a negative number. *We do not know which of these three cases occurs.* Now Brouwer thinks that we are dealing here with a real number which is neither positive nor zero nor negative. And through this result he explodes a whole series of famous propositions of classical mathematics.

The reader will naturally say that a distinction should be drawn between stating that the equation has a solution and that we know it. Surely only one of the three cases can occur, namely, the sequence either tends to 0 or it stops at a positive or negative value. The fact that we cannot settle this matter has nothing to do with its substance. Brouwer would reply that possibly the problem is *undecidable in principle* and that one has therefore no right to argue in this way.

If *we* had to take a stand on this, we would say that if Fermat's problem is undecidable, then this sequence does not represent a real number at all. For, the essence of a real number is actually its comparability to the rational numbers—it is only for this reason that it can be indicated as a point on a straight line. If there are number-like formations which cannot be compared with the rational numbers, we have no

13. THE REAL NUMBERS

right to insert them between the rationals, and it is only logical to deny the title of real numbers to such formations.

The very same consideration can be applied to decimal fractions by making the sequence of numerals of a decimal fraction depend in some way on the solution of a mathematical problem. An example of this would be the decimal fraction $0 \cdot a_1 a_1 a_2 a_3 \ldots$, where the numeral a_n shall be equal to 1 or equal to 0 according to whether Fermat's equation $x^n + y^n = z^n$ has a solution for n or not. The decimal fraction would therefore begin 0.11000 . . ., but we would not know today whether it is $= 0.11$ or > 0.11. We would again say that if the problem cannot be solved, then a real number is not defined by this prescription.

These considerations make us want to trace the relation between real numbers and decimal fractions. One usually believes that a decimal fraction is determined, defined, given, if its numerals are known, that is to say, if the sequence of numerals is written out to infinity. Strictly speaking, however, is this correct? Isn't there a process working behind the sequence of numerals, namely, the *law* which generates the numerals? The question could also be stated thus: Are decimal fractions conceivable which are without rule and infinite? Are they real numbers at all? Suppose, for example, that the succession of the numerals was determined by lot, would a real number be thereby defined—yes or no? This question would probably be answered today in different ways. *We* would say: even if the interval were narrowed down more and more by continuing the process, we would still not have the right to speak here of a number; this is because such formations satisfy laws different from those valid for real numbers. (Can one ask whether such a decimal fraction is rational or irrational?) What leads us astray in this regard is an analogy with visual space. Here if a segment is made smaller, it gradually goes over into a point, and similarly we visualize the approximation of a real number. If the sequence of numerals, so one thinks, is merely continued on and on, then the interval approaches a point. This conclusion is false. No matter how far the sequence is carried out *the position of the point is always just as undetermined as ever*. It would be different if we knew a law; for then this law would itself be the real number. We thereby arrive at the source of the entire deception, namely, the confusion between *extension* and *law*. One imagines that an infinite decimal fraction could be determined in two ways: by enumerating the numerals *or* by a law. However, it is not so. A real number *generates* extensions, it is not an extension.

It would be interesting, from this point of view, to look at certain theories of probability which rest on the assumption that a limit can be defined by an empirical process, for example, by a statistical series. If the reader wishes to learn something more about this, he may consult an article by the author in *Erkenntnis (Knowledge)* of 1930.

To summarize, we can say that the real numbers were created in order that the limit operation could be performed without restriction. Only, this operation is not as simple, clear and transparent as, let us say, the operation of subtraction; instead, it is a rule, a law for the generation of rational numbers. Now, there are very different types of laws, and therefore there will also be various *kinds* of real numbers. At the beginning it looks as if they formed a *uniform* system, like the rationals, and this view is underlined by the common notation, as decimal fractions. But the very discovery of Gödel shows that altogether different relations exist in this regard. Namely, for every arithmetic real numbers can be found which cannot be defined by the methods of this arithmetic. Before a keener look the system of real numbers is resolved into an infinite set of different systems which have only a certain similarity to one another.

Finally we ask: is the consistency of the theory of real numbers a consequence of their construction from the rational

13. THE REAL NUMBERS

numbers? Formerly, this was believed to be true. The definitions for operating with sequences, as we have seen, are exact copies of certain theorems (cf. Chapter 11); consequently, it was assumed that every contradiction between our conventions must entail one of the same type between demonstrable theorems. The investigations of Gödel have shown, however, that this is not so; on the contrary, the consistency of the theory of real numbers cannot be demonstrated from the standpoint of the natural numbers.

The deeper reason why the process of reduction breaks down at this point can be made clear as follows: whereas a statement regarding a positive or negative or a rational number can be entirely transformed into a statement about natural numbers, this is no longer possible in the case of real numbers. Here, an essentially new element is introduced, namely, the concept of *law* (in the case of Cantor's sequences of numbers), respectively, the concept of *class* or property (in the case of Dedekind's cuts). This means that the calculus with real numbers is a calculus with laws or classes of rational numbers and therefore cannot be translated into the language of rational numbers.

14. Ultrareal Numbers

There are problems which cause us to go beyond the real numbers, namely, to number systems which no longer satisfy our four requirements. To begin with, we will consider two examples.

1. If the reader represents the function $y = \frac{1}{x}$ graphically —in doing this he can restrict himself to the positive values of x— he will find that the curve becomes steeper and steeper toward the origin and, as x converges to 0, grows beyond all bounds. Such a point is called a point of infinity or a pole. Let us now consider various functions, each of which may have a pole at the point $x = x_0$. We are then confronted with the following question: can the rate of growth of these functions be compared to each other? Can we grade the process of becoming infinite? Let us consider, for example, the functions $f(x) = \frac{1}{x}$ and $g(x) = \frac{1}{x^2}$. If the reader draws the graph of the latter function he will notice that it ascends steeper than the graph of the first function. More specifically, the ratio of the two functions $\frac{g(x)}{f(x)}$ goes to ∞. In this case the function $g(x)$ is said to approach infinity of a higher order (at a sharper, more rapid rate) than the function $f(x)$, or, more briefly, infinitarily $g(x)$ is greater than $f(x)$. If, however, the ratio $\frac{g(x)}{f(x)}$ tends to a fixed numerical limit (neither 0 nor ∞), the two functions are said to be of the same order of infinity, or infinitarily equal.

14. ULTRAREAL NUMBERS

Since the relations "greater," "equal," "smaller" form an exhaustive disjunction, the poles can be sequentially arranged. The question raised above can now be made more precise. Can a measure be introduced for the rate of becoming infinite? Can the individual poles be characterized by numbers so that the greater number always corresponds to a pole greater infinitarily? It is easy to set up an unbounded sequence of functions each of which ascends more rapidly to infinity than the preceding; for example, the functions $\frac{1}{x}, \frac{1}{x^2}, \frac{1}{x^3}, \cdots$ Further terms can be inserted into this sequence by also taking fractional and, finally, real exponents. We will now try to assign the number a as the "measure" or "degree" of the function $\frac{1}{x^a}$. *In designating the poles of this particular type of function in such a way, all real numbers* (from 0 on) *are used up.* But there are still other functions which become infinite at the point $x = 0$, for example, $|\log x|$. On comparing $|\log x|$ with any one of the functions $\frac{1}{x^a}$, a calculation shows that $|\log x|$ goes to infinity more slowly than any function of this kind. $|\log x|$ *has a lower* rate of increase than all functions of the form $\frac{1}{x^a}$. A function such as $\frac{|\log x|}{x}$ will therefore be infinitarily greater than $\frac{1}{x}$ and still smaller than $\frac{1}{x^{1+\varepsilon}}$, no matter how small ε is chosen. If a measure is to be assigned to this function, we can choose for this neither the number 1 nor any number > 1—*the scale of the real numbers is no longer sufficient to designate the order of poles*. A denser structure belongs to the sequence of poles than to the continuum of real numbers. In order to grade them, we must have recourse to an "ultrareal" number system—to be sure, however, at the price of dropping some of the requirements which we have placed on the number concept, say, the Archimedean axiom. The reader may notice that in this problem he is concerned, strictly speaking, only with the discovery of a system of *ordinal numbers*—for we have no absolute need to define operations for calculating with these numbers.

2. In many considerations the "horn-shaped angles" are of interest; these were already known and discussed in antiquity. If two curves intersect, the angle between them is generally taken as the angle formed by the tangents that are drawn at the intersection point. These angles form an ordinary Archimedean system of quantities. Now another standpoint can be taken just as well. Intuition suggests that we should fix our attention on the horn-shaped space between the two curved lines, the shape itself, and ask whether these spaces cannot be directly compared to one another. This is certainly possible. First of all, we can split the horn-shaped angle by a straight line into two parts (cf. Fig. 25). Hence we restrict ourselves to angles which have a straight line as one side. It is natural then to define an angle as being greater than another if it extends beyond the other (provided that the two angles are placed one on the other such that their vertices and straight-line sides coincide). It is clear that by this determination the horn-shaped angles form an ordered system.

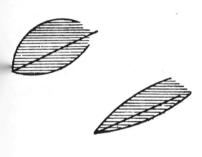

Figure 25

If we apply this in particular to the angle which is formed by a straight line and a circle tangent to it (this angle is equal to 0 if measured in the usual sense), then the circle with the greater radius is associated with the smaller horn-shaped angle (cf. Fig. 26). We will therefore set the reciprocal value of the radius $\frac{1}{r} = \omega$ as the measure of the magnitude of the curvilinear angle. It is clear that the sequence $\omega, 2\omega, 3\omega, \ldots$ can be

221

formed by shrinking the radius in the appropriate ratio. However, no matter how far this sequence may be carried out, ω will always remain smaller than the straight-line angle a.

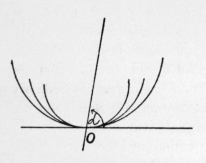

Figure 26

Consequently, ω is actual-infinitesimal relative to a, and this shows very clearly that the Archimedean axiom breaks down in this case.

If these things are to be computed out, we must first of all find an expression for the fact that the angles considered above are actual-infinitesimal relative to an ordinary angle. This will now be done by introducing an actual-infinitesimal unit η. It is defined as the angle which the circle of radius 1 forms with its own tangent.

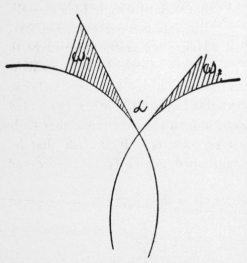

Figure 27

Any other circle will then form with its tangents the angle $\omega = \frac{1}{r} \cdot \eta$. Now, if two circles intersect, the horn-shaped angle is composed of three portions: the angle a which is formed by the two tangents, and two actual-infinitesimal angles ω_1 and ω_2.

Accordingly, the measure of the total angle is defined as the sum

$$a + \omega_1 + \omega_2$$

or

$$a + \frac{\eta}{r_1} + \frac{\eta}{r_2},$$

where r_1 and r_2 designat[e]

the radii of the two circles. The quantity η is thereby characterized by $n \cdot \eta < 1$ no matter how large n is chosen.

To transfer this definition to arbitrary curves, the curves are first replaced at the intersection point by their "osculating circles" (that is, by those circles which hug a curve as closely as possible at the given position); and then the convention is made that the angle between the curved arcs is to be the angle between the corresponding circular arcs. It is thereby assumed that the curves possess a "curvature"; this is no more self-evident than that they have a direction.

If the abstract formalism is stripped from all this, it becomes a question of erecting a calculus with elements of the form $a + b\eta$ in which the Archimedean axiom is no longer valid. Calculating rules must now be established for these symbols. On this occasion the question naturally arises as to the meaning of $\eta \cdot \eta^2 \eta^3$ etc. One suggestion is to interpret them as actual-infinitesimal quantities of higher order. This standpoint is taken by Veronese in his work on the foundations of geometry. He thinks of them as forming a hierarchy of actual-infinitesimal quantities each of which is actual-infinitesimal relative to the preceding one. Then he constructs a non-Archimedean geometry out of them in which the individual points on the axis of abscissas are given by an expression of the form

$$a + b\eta + c\zeta + \ldots$$

where a, b, c, ... indicate ordinary real numbers. What is described as a point in the usual geometry, here splits up, so to speak, into a point world, just as a bright spot in the sky is resolved into a star cloud by a telescope.

The construction of such number systems is important for those investigations about the foundations of geometry which seek to settle the question, among others, of the independence of the axioms. This is the question whether, among the axioms which are taken as a basis for a definite geometry, theorems are not present in a disguised form whose proof has not been found

14. ULTRAREAL NUMBERS

up to now. This is how the matter stands: If a consistent geometry can be constructed in which all axioms are valid with the exception of one, then the latter axiom can certainly not be a logical consequence of the remaining ones. Strictly speaking, the only purpose in constructing a non-Archimedean geometry is to demonstrate the independence of the Archimedean axiom from the others.

The reader can now judge for himself the view that to deal with non-Archimedean systems of quantities is nothing but idle play. The reason why such numbers seem unusual and strange to us seems to be that they have nothing to do with the concept of extensive quantities, which we mainly think of in this connection. However, just from the standpoint of logic it is not self-evident that the various quantities of one kind, for example, lengths, can be compared. "This insight," says Hilbert "is, as you know, of essential significance in geometry, but it seems to me of essential interest in physics, too, for it leads us to the following result: the fact that we can find the dimensions and distances of bodies in the universe by merely joining terrestrial distances, that is, that astronomical lengths can be measured by terrestrial measurements, and the fact that distances within the atom can be expressed by metric measure, are in no way merely a logical consequence of propositions about congruent triangles and geometrical configurations, but only a result of empirical research. The validity of the Archimedean axiom in nature simply requires, in the designated sense, the confirmation by experiment just as, let us say, the proposition regarding the sum of the angles in the triangle in the known sense."[1] It would be interesting to imagine experiences which would force us to surrender the Archimedean axiom.

It has already been mentioned that the intuitive continuum is not the mathematical one. The real numbers cannot be properly applied to the former. Some more examples may illustrate

[1] *Axiomatisches Denken* (*Axiomatic Thinking*), Math. Ann., 78.

this further. Let us think of seeing a large number of points on a graph which are so clustered that they produce approximately the picture of a strip, about as it would appear to an eye viewing it from some distance. This strip is supposed to be nowhere sharply defined, but to disappear gradually in the surroundings. Now, how wide is the strip? Since it cannot be sharply defined, we are dealing here with a measure in an entirely different sense, rather with the rendering of the *impression* made by the strip, namely, with what is called an *estimation*. (This already follows from the fact that two numerical statements involving numbers which are not too far apart in magnitude need no longer contradict one another, whereas they would be contradictory in the case of the ordinary kind of measure.) A similar question is the following: where does the plateau begin in this mountain profile? Here, too, it is not a question of marking this point sharply. We find ourselves in an analogous situation if we ask: where does the sound come from? We could reply by pointing in the direction, but only approximately; this indicates that we are concerned here with another concept of direction or with another mensuration of angles. An analogous question is the following: is this orange color *exactly* halfway between red and yellow? It has no sense to speak of "exact," unless a method of measurement is indicated. (Color composition.) But then we are already moving in the domain of physics and not of immediate intuition.

In all these cases the use of real numbers is not entirely appropriate. Perhaps a mathematics of the vagueness will some day be constructed which is better adapted to such relations. Thus, we could think of axioms being formulated for numbers according to which "greater," "equal," "smaller" would not form an exhaustive disjunction; instead there would be provision made for peculiar zones of indeterminancy or of poor discernibility. But finally this would only amount to formulating the grammar of the word "approximately."

15. Complex and Hypercomplex Numbers

There is a much older extension of the number system, namely, the imaginary numbers, which appeared in occidental mathematics about the middle of the 16th century. Two motives forced such an extension. One is algebraic in nature. The non-solvability of the equation $x^2 + 1 = 0$ was the occasion for admitting a new type of number in calculations, namely, the square roots of negative quantities. Besides, a second motive should not be overlooked, which came forward only later. With the emergence in geometry and in physics of the concept of "directed quantity" or vector, it became desirable to form a purely arithmetical analogue to the concept of vector. This provided the stimulus for the creation of higher complex numbers.

For the present, however, we wish to speak of the ordinary complex numbers, and in this regard we must first say something about the significance of imaginary numbers for the whole of mathematics. This significance is not due only to the fact that they make it possible to extract the roots of negative numbers. Rather, by introducing these numbers algebra and analysis experience a tremendous simplification. The true value of such numbers lies in the fact that they enable us to form connections between entirely different parts of mathematics. Furthermore, a complete insight into many problems is attained only by doing away with the restriction to the reals and entering the complex domain. As a first example, we cite here Euler's famous relation:

15. COMPLEX AND HYPERCOMPLEX NUMBERS

$$e^{ix} = \cos x + i \sin x$$

which all at once discloses a connection between functions that seem to be of entirely different types. The complete development of this idea is found in the modern theory of functions, where dependences between complex numbers, namely, functions

$$w = f(z),$$

where $z = x + iy$ and $w = u + iv$ are complex, are studied quite generally. The geometrical interpretation now consists in imagining two planes, the z-plane and the w-plane; every point of the z-plane can be made to correspond to a point in the w-plane by the calculating rule f. Each such function $w = f(z)$ produces, therefore, a mapping of one plane on the other, and it now turns out that the essential, deeper-lying properties of functions manifest themselves in the very kind of mapping effected by them. Thus, very large classes of functions can be characterized by startlingly simple properties, for example, by uniqueness, occurrence of poles, etc.

Many phenomena cannot be understood at all from the standpoint of real numbers, for example, the behavior of the function log x. On asking about the logarithm of a negative number, the following consideration presents itself. Let us set $\log(-1) = x$. Then $\log(-1)^2 = 2\log(-1) = 2x$. On the other hand, however, $\log(-1)^2 = \log 1 = 0$. Hence $2x = 0$ which amounts to saying that the logarithm of -1 is zero — which is obviously incorrect. The source of the error cannot be discovered by the methods of school mathematics. Everything is cleared up as soon as one rises to the standpoint of function theory. It then turns out that the function log z is infinitely many-valued, that it represents, so to speak, infinitely many individual functions, which are, however, again connected in a very definite way. Only thus can an insight into the complicated organism of this function be obtained. If we restrict ourselves to the portion given by the reals, then the vital point escapes us; we resemble the observer in the cave simile of Plato, who sees only the shadows of objects as they pass by, and to whom the actual entities remain eternally strange. Probably Gauss also had this in mind when he said that for him "analysis is a self-dependent science, which by disregarding those fictitious quantities would lose extremely in beauty and fullness, being forced all the time to add the most burdensome restrictions to truths which otherwise are universally valid."

The geometrical visualization of complex numbers, which is due to Gauss, has been very useful in mathematics, but it has not justified the introduction of imaginary numbers. In 1835 Hamilton developed a theory from a point of view which is in vogue today. He interpreted a complex number as a couple of real numbers, whose laws of combination can be arbitrarily chosen. Let us designate such a number couple by (a, b) and call a and b the components. The following sketch will bring the main points of the structure into prominence.

DEF. 1. Two number couples are said to be equal if their components are equal; in symbols:

$$(a, b) = (c, d) \text{ if } a = c \text{ and } b = d.$$

The concepts "greater" and "smaller" are not introduced.

DEF. 2. The sum (difference) of two number couples is the couple formed from the sum (difference) of the components:

$$(a, b) \pm (c, d) = (a \pm c, b \pm d).$$

It is readily seen that all formal conditions for these concepts are satisfied.

DEF. 3. $\quad -(a, b) = (-a, -b).$

From the definition of sum it follows that

$$(a, b) + (a, b) = (2a, 2b),$$

which we can briefly write as
$$2(a, b) = (2a, 2b).$$
By induction we recognize that in general
$$n(a, b) = (na, nb).$$
Hence, a number couple is multiplied by an integer by multiplying its components by this integer. The division by an integer is thereby explained, too. Thus, if we set
$$\frac{1}{m}(a, b) = (x, y),$$
then $\quad (a, b) = m(x, y) = (mx, my),$

which yields $x = \frac{a}{m}$, $y = \frac{b}{m}$. From this it follows further that
$$\frac{n}{m}(a, b) = (n \cdot \frac{1}{m})(a, b) = (\frac{n}{m}a, \frac{n}{m}b).$$
True, a new assumption has been made in doing this, since the associative law of multiplication must be valid. By Def. 3 the last formula is still correct if $\frac{n}{m}$ is negative so that for any rational number r we have in general
$$r(a, b) = (ra, rb).$$
On the other hand, it cannot be proved that for any arbitrary real number ϱ
$$\varrho(a, b) = (\varrho a, \varrho b)$$
We will take it as a convention agreed upon so as to conform with the multiplication with rational numbers, hence, to satisfy the postulate of permanence.

We are now in a position to represent every number couple in a normal form. Namely, we have:
$$(a, b) = (a, 0) + (0, b) = a \cdot (1, 0) + b \cdot (0, 1).$$
The number couples $(1, 0)$ and $(0, 1)$ appearing here will be called the "complex units" and will be designated by e_1, e_2. Then, every number couple can be represented as a linear combination of the units

15. Complex and Hypercomplex Numbers

$$(a, b) = ae_1 + be_2.$$

What shall we now understand by the product of two number couples? If we write $(a, b) \cdot (c, d)$ in the form

$$(ae_1 + be_2) \cdot (ce_1 + de_2)$$

and calculate this product by the usual rules, namely, using the distributive law, then four different combinations of the units appear, which we must define. If the multiplication is not to lead us out of the system, then $e_r \cdot e_s$ must again be a number of the system; this means that it must be a linear combination of the units $\lambda_1 e_1 + \lambda_2 e_2$. Among the infinitely many definitions presenting themselves one must now be chosen. Our only guiding star is the purpose which the system is to fulfill. This purpose is attained if we make the following conventions:

$$e_1 \cdot e_1 = e_1$$
$$e_1 \cdot e_2 = e_2 \cdot e_1 = e_2$$
$$e_2 \cdot e_2 = -e_1.$$

Through this set of formulae one complex system of number is singled out among the infinitely many possible ones. Multiplication then assumes the following form:

$$(a, b) \cdot (c, d) = (ac - bd, ad + bc).$$

It is now easy to see that the number couples $(1, 0)$ and $(0, 0)$ play in this system the same role as 1 and 0; thus, for example,

$$(a, b) + (0, 0) = (a, b) \text{ just as } a + 0 = a.$$
$$(a, b) \cdot (0, 0) = (0, 0) \text{ just as } a \cdot 0 = 0.$$
$$(a, b) \cdot (1, 0) = (a, b) \text{ just as } a \cdot 1 = a.$$

In general the number couple $(a, 0)$ will correspond to the real number a, and we recognize that *there is a subsystem the complex numbers*, the numbers of the form $(a, 0)$, which is associated *to the system of real numbers by a one-to-one isomorphism*. (We cannot speak at the present of similarity because we have not defined the relations "greater," "smaller."

In order to pass over to the usual representation we only have to set $e_1 = 1$ and $e_2 = i$. We thereby obtain the multiplication rules

$$1 \cdot 1 = 1$$
$$1 \cdot i = i \cdot 1 = i$$
$$i \cdot i = -1.$$

This last formula is the basis of the fact that the equation $x^2 = -1$ can be solved by means of this system. If we had set up the multiplication rules otherwise — which from the standpoint of logic would be just as permissible — we would not have attained this purpose. Hence it is the application we may wish to make which is the decisive factor in the choice of definition.

Division is introduced as the inverse operation of multiplication. On setting

$$\frac{1}{a+bi} = x + iy,$$

we have

$$1 = (a + bi) \cdot (x + iy),$$

which means

$$1 = (ax - by) + i(ay + bx).$$

On comparing reals with reals, imaginaries with imaginaries, we obtain

$$x = \frac{a}{a^2 + b^2}, \quad y = -\frac{b}{a^2 + b^2}$$

and therefore

$$\frac{1}{a+bi} = \frac{a}{a^2 + b^2} - \frac{b}{a^2 + b^2} i.$$

We have not attempted up to now to order these quantities. We now wish to make a convention by defining: (a, b) shall be greater or smaller than (c, d) according as $a > c$ or $a < c$; however, if $a = c$, then the decision shall depend on whether $b > d$ or $b < d$. In the sense of this convention, we have

15. COMPLEX AND HYPERCOMPLEX NUMBERS

$$(0, 1) < (1, 0)$$

However, no matter how often the left number couple is multiplied by an integer, it will always be true that

$$n \cdot (0, 1) = (0, n) < (1, 0).$$

Hence the Archimedean axiom is not valid (which agrees with the general discussions of p. 209). In this interpretation $(0, 1)$ appears as actual-infinitesimal relative to $(1, 0)$.

Is it not possible to introduce numbers whose representation requires the three-dimensional space, more generally, an n-dimensional space? For this purpose the n-tuples of numbers

$$(a_1, a_2, \ldots a_n)$$

have to be considered, and the definitions of equality, sum, difference, and multiplication by a real number have to be given for them in an entirely analogous manner. Any such expression can then be represented as a linear combination of n "units":

$$(a_1, a_2, \ldots a_n) = a_1 e_1 + a_2 e_2 + \ldots + a_n e_n.$$

If we also wish to introduce multiplication, and let ourselves be guided by analogy with the ordinary calculations with letters, we have to come to an agreement regarding the product of two units $e_r \cdot e_s$. There are two possibilities. Either this product is not representable by the means hitherto applied. In this case perhaps new units must be introduced, whose multiplication by the old numbers again requires new units, so that the original system must be extended still more. Or else, the system may be closed in the sense that the product of two units is again a number of the system; in this case, we are led to the formulation

$$e_r \cdot e_s = a_1 e_1 + a_2 e_2 + \ldots + a_n e_n.$$

For each of the n^2 possible combinations we must now think of such a formula being formed; and *the characteristic of a*

15. Complex and Hypercomplex Numbers

such number system lies in the specification of the n *coefficients of these formulae.*[1]

The study of these "hypercomplex" numbers has brought to light certain general truths. Weierstrass, in a lecture given in Berlin 1863, showed that such number systems are actually conceivable, but that in these systems certain fundamental calculating laws must be renounced. Either, in calculating with these numbers the commutative law of multiplication may not be valid, so that a · b is distinct from b · a (this would yield the result that there would be two kinds of divisions in such a system). Or else, the commutative law may be retained, but some other important law of arithmetic may be lost, for example, the proposition that a product of two numbers can vanish only if at least one of the two factors is zero; in this case, division may become infinitely ambiguous. The permanence principle leaves us in the lurch here, since it no longer uniquely points out the way we have to take for the extension. If we require that the commutative law of multiplication remain intact and that an algebraic equation (for example, $ax + b = 0$) with coefficients distinct from zero should not have infinitely many solutions, then the system of ordinary complex numbers is the only one left. Hence this system occupies a preferred position. Therein lies the answer to the question whose solution Gauss had announced but had not given, "why the relation between entities representing a manifold of more than two dimensions cannot yield other kinds of quantities admissible in general arithmetic."

The reader will now ask: Isn't all of this merely idle play? Can one begin anything sensible with such hypercomplex numbers? Here we will only point to such a system, which has attained a certain application, namely, the *quaternions* of

---

[1] Grassmann, one of the creators of higher complex numbers, considered no less than 16 distinct kinds of multiplications in a publication during the year 1855.

233

15. Complex and Hypercomplex Numbers

Hamilton.[2] These are, as the name implies, four-term numbers constructed out of one real and three other units, which can be interpreted as directed quantities (vectors) in our three-dimensional space. Without going into these things in greater detail, we only wish to say that the quaternions played a very useful role in the mathematical treatment of rotation (more precisely: of rotary extension, namely, of a rotation of three-dimensional space about the origin combined with an extension in a certain ratio) and that their importance in physics is based on this fact. Such rotary extensions play a role in the very interpretation of certain formulae of the theory of relativity—the "Lorentz transformations"—in the four-dimensional space-time world of Minkowski. From the papers left by Gauss it is to be seen that he was already in 1819 acquainted with quaternions and their applications.

---

[2] *Lectures on Quaternions*, 1853.

16. Inventing or Discovering?

Our discussions are a good testing ground for the teachings of school logic. According to these teachings, the formation of concepts is attained through *abstraction* (separation of attributes) or through *determination* (addition of attributes). Is this the way it actually takes place? Is it true that at first we have a general concept of number, and that this concept is then narrowed down stepwise by the addition of formative attributes? For example, do we descend, first, to the concept of complex number, then, to that of real number, etc.? Or are we to imagine, conversely, that the general concept of number is obtained by starting from the concepts of the number types and then calling attention to the familiar properties? But, what are these general properties? The scheme delineated earlier (on p. 209) is too wide for the natural numbers and integers, and too narrow to take in the complex numbers and those of Veronese. And this is a very characteristic situation; namely, whatever the system of conditions we set up, we are never sure of having defined exactly the concept of number. For, what guarantee do we have that new number types will not be discovered which violate our conditions? Or shall we declare in this case that such concepts should not be called numbers?

"Cardinal number," "integer," "rational number" will be called sharply defined concepts, for each of them is defined by a calculus. However, what are we to understand by a number (in general)? The best answer that can be given is to declare that the above formations as well as all those which are somehow similar to these fall under the concept "number"; wherein we deliberately say nothing about the kind of similarity. If one thinks that in mathematics all concepts must be clearly and sharply defined, we will only point out that the mathematician has no need of the general concept of number. In fact, where did this concept ever occur in our proofs?

In regard to the question: what is a number? the situation is the same as in the question: what is a point? First of all, does the word "point," in the Euclidean geometry of the plane, have the same meaning as in the Euclidean geometry of space? By no means; a point satisfies more rules in the latter case than in the former. Hence this word has a clear-cut meaning only in a definite geometry. If we place metric geometry, affine geometry, projective geometry, and topology side by side, each of them attaches a different meaning to the same word; strictly speaking, the meaning is characterized only by the enumeration of all the axioms of the science under consideration. If the word "point" plain and simple is used without further explanation, supposedly, that is, analogous to the usage of the word point in the Euclidean geometry of the plane, then it loses its precise meaning and assumes a vague, indistinct one. Similarly, the concept of number becomes suddenly vague and indistinct as soon as we no longer allow it to be defined by definite calculus. It then becomes rather uncertain what is still to be called a number. For example, an algebraic equation or an element of an abstract group could each be called "number"; or else, a sentence with which symbolic logic calculates (indeed, certain analogies speak in favor of that, since in the case of sentences, certain combinations are defined which are called "sum" and "product," and furthermore the role of tautology and contradiction can be compared to those of the numbers 0 and 1). Whether the concept of number is to be widened to this extent is, at long last, a question of feeling and tradition.

Let us put it this way: the individual number concepts (cardinal number, integer, etc.) form a *family*, whose terms have a family similarity. What does the similarity of the members of a family consist in? Well, some have the same nose, others the same eyebrows and still others the same gait; and these similarities overlap to some extent. We do not have to maintain that they all must have a property in common; even if there were such a property, it need not be the one that constitutes the family similarity. We will describe this by saying that the word "number" does not designate a concept (in the sense of school logic), but a "family of concepts." By this we mean that the individual number types are related to each other in many ways even though they may not have one property or one trait in common.

This is also valid of the expressions "arithmetic," "geometry," "calculus," "operation," "proof," "problem," etc. They all designate families of concepts, and it is of little value to start a controversy regarding their exact definition. In wishing to explain the concept of arithmetic we will point to examples and allow the concept to reach as far as the similarity reaches in these examples. The very openness, non-closure, of these concepts also has its good points, for it gives language the freedom to comprise new discoveries in a known scheme.

These things had to be discussed here because they provide the background for a question which appears again and again to the meditative mind, namely, are the numbers creations of the human mind, or do they have an autonomous kind of being all of their own? *Are they invented or are they discovered?* A person who looks over the foregoing considerations cannot be in doubt as to the answer. If we were to state our view in a brief formula, we would say: the meaning of a symbol follows from its application. The rules of application only

16. INVENTING OR DISCOVERING

impart to the symbols their meaning. Thereby, we reject the interpretation that the rules follow from the meanings of the symbols. It is this very interpretation that is a favorite view of many philosophers; we will examine this opinion more closely to see what value it may then still hold.

Frege is the most important representative of the above interpretation. He sought to reduce *ad absurdum* the "formal interpretation" of arithmetic — as he called the axiomatic method of reasoning coming up at that time — by a penetrating analysis of its assumptions. His arguments are aimed in part at us, too, and therefore it will make things clearer if we measure our view against the strength of his reasons. His interpretation can be summarized into four arguments.

1st ARGUMENT. Arithmetic can, it is true, be considered as a game with symbols; then, however, the proper sense of the whole is lost. In giving anyone certain rules for the use of the equality sign — namely, telling him that he can go from the formula $a = b$ over to the formula $b = a$, furthermore, from the formulae $a = b$ and $b = c$ to the formula $a = c$ — have I thereby communicated to him the *sense* of this symbol? Does he now understand what the symbol "$=$" means? Or have I only given him a direction for using the symbol mechanically, which he could also follow without any idea of its sense? The latter is certainly true. But then the formal interpretation of arithmetic loses what is most important, namely the sense which expresses itself in the symbols. This sense can be comprehended only through thinking, through a mental process.

REPLY. Let us assume that such is the case. Why not then describe this mental process right away? But, if someone asks me what the formula "$1 + 1 = 2$" means, I will not answer by describing my state of mind; instead, I will answer with an *explanation of the symbols*. I will say that this formula means in words "One and one is equal to two" or " '$1 + 1$' ca be replaced by '2' "; or I will illustrate the use of this formula by an example. Hence I give as an answer 1. translations of the arithmetical formula into word language, and 2. applications. I combine this symbol with other symbols, make it a part of a system of symbols and operations, and *this* gives it its sense.

But, one may say, I certainly know what the symbol "$=$" means. And to the question "Well, what does it mean?" I may give a series of answers such as: It means "equal"; it means "replaceable by"; if it stands between two symbols, it means that the left is the same as the right; for example, 2×2 is equal to 4, $(a + b)^2$ is equal to $a^2 + 2ab + b^2$, briefly, we symbolize the various applications of the symbol. Not *one* answer has been given but many. The meaning of the formula $1 + 1 = 2$ is, as it were, the whole aura on the background of our word language, which is again a web of symbols and operations.

This discussion shows at the same time the justified core of Frege's criticism. Namely, by concentrating exclusively on the formal side of arithmetic, by detaching arithmetic from its applications, by cutting through all the strings which tie arithmetic to our word language, we certainly obtain a mere game. If a child is taught only such formulae and nothing else, anything we call the sense of the whole matter would escape him. Frege was therefore right in seeing *more* in arithmetic than such a game of formulae. However, what is missing here is not a process of understanding which accompanies the reading of the formulae, but the interpretation of the formulae. And this interpretation consists in exactly nothing else but in integrating the rules of the calculus in a wider syntactical connection. If I teach a child along with the formulae also the translations of these formulae into words and various examples of their application—will the proper sense elude him even then? And will he still be making a merely mechanical use of the symbols?

2nd ARGUMENT. It is, therefore, application which distinguishes arithmetic from a game. However, what does application depend on? "Without a conceptual content," says Frege, "an application is not possible either. Why can't an application of a position of chessmen be made? Evidently, because it expresses no thoughts. Why can applications of arithmetical equations be made? Only because they express thoughts." (*Grundgesetze der Arithmetik* [*Basic Laws of Arithmetic*] Vol. II, Section 91.)

REPLY. Let us imagine an arithmetical game has been invented which looks exactly like the usual arithmetic except that it will never be applied; instead, it is only used for amusement — would it still express some conceptual content? Most people would answer this question negatively. What then, must be added to an equation of arithmetic if it is to express a conceptual content? Application, and that is all. It is mathematics if the equation is used for the passage from one proposition to another (cf. Chap. 9); otherwise, it is a game. To say that a position of chessmen expresses no thought is hasty, for this depends entirely on us. Suppose the troops in battle move as the men on the chessboard, then this could induce us to express a sense through the position of the men; a chess move would now receive a meaning, and the officers would bend over the chessboard just as they do now over general staff maps. The motion of the men would just be an image of actual proceeding and not a "mere game." "Because a chess move expresses no thought, it cannot be applied." Wouldn't it be more correct to say: "Because we have not provided for an application, a chess move expresses no thoughts"?

An opponent will perhaps reply that the statement just proves the correctness of Frege's views. For, why is it that a position of chessmen can express something? Certainly only because the men on the board signify troops on the battlefield; hence, because they are *symbols of something*. This leads us into the range of ideas of the

3rd ARGUMENT. The application of arithmetic is founded on the fact, "that the number symbols mean something, whereas the chessmen do not" (Section 90). Don't say the mathematician creates numbers by his definition. "Here the point is to be clear about what defining means and what can be attained by it. Frequently, we seem to credit him with having a creative power, whereas he can actually only succeed in making something prominent through a definition and giving it a name. Just as the geographer does not create a sea when he draws boundary lines and says: I wish to call the Yellow Sea the part of the water area bounded by these lines, so also the mathematician, strictly speaking, cannot create anything through his ability to define. Nor can we, by mere definition, endow a thing, as if by witchcraft, with a property it simply has not got, except the one to bear the name we have given it. But I can only consider it as a scientific superstition that an oval figure drawn with ink on paper shall receive through a definition the property of yielding one if added to one. We could just as well make a lazy student into an industrious one by a mere definition. Not until it has been proved that there is one and only one object with the required property, are we in the position to assign the proper name 'zero' to this object. To create zero is therefore impossible" (Vol. I, p. XIII).

Hence this argument says that a symbol must designate something, otherwise it is merely printer's ink on paper. Arithmetic is a science only because numbers exist.

REPLY. The essential point of the matter lies in the last statement. We wish neither to attack it nor to agree to it. Instead, we simply ask, what sense can be attributed to this statement? From this will follow what we have to think of it. That numbers are not the same as the symbols we write on paper, that it all rather depends on the use we make of the symbols, is so clear that nothing more need be said about this. However, Frege wishes to say more than this, in fact much more. He thinks that numbers are already there somehow, so that the discovery of imaginary numbers is comparable, let us say, to

16. INVENTING OR DISCOVERING

the discovery of an unknown continent. What shall we think of this interpretation? Let us illustrate the matter by an example, let us assume that we have invented a number system consisting of n units $i_1, i_2 \ldots i_n$ and have given rules for calculating with these numbers. We now ask whether these numbers exist or not? Frege must answer this question negatively and has done so. He expressly writes regarding this example: "It is nowhere proved that there are such units, it is nowhere proved that one has the right to create them. It is impossible to regard ,i_1', ,i_2' etc. as meaningful proper names, similarly to '2' and '3'" (Vol. II, Section 141). He would have granted at most that it was an interesting game. But, what if this game proves to be uncommonly fruitful for mathematics? If they enable us to solve problems which were unassailable up to then, shall we still continue to say that these numbers do not exist in actuality? No mathematician will hold such a view. Instead, one will work with these formations just as with negative or irrational numbers. It must not be said that these are empty possibilities. For, this is how things actually happened with the introduction of imaginary numbers, of Hamilton's quaternions, of the actual infinitesimal numbers of Veronese; all of them were originally no more than elements of a game, until their great usefulness was uncovered. What is more natural than to say: in the case of imaginary numbers, quaternions, etc. an application has been found, and for that reason are they a subject matter of science in other cases, such an application has failed to appear, and therefore it is a game?

A follower of Frege would perhaps reply: this just shows that, in the one case, something objective presents itself, while in the other, this is not so. Very well! But then, at any rate, still remains true that the criterion for existence is the applicability. As things now stand, the entire *sense* of the statement which confers objective existence on any number is incident to the applicability, and this statement means not one iota more

4th ARGUMENT. If new numbers could actually be created in order to solve a problem which was unsolvable up to then — for example, to impart a solution to the equation $x^2 + 1 = 0$ — why isn't this simple method employed at all times to rid oneself of unsolvable problems? For example, the equation $1^x = 2$ does not have a solution, at least as long as one is restricted to the numbers known up to now. Very well, let us create a new number, and now this equation is solvable. Is this legitimate? No, the word "create" won't do. In the problem to solve the equation $x^2 + 1 = 0$, an extension of the number domain was possible; in the case of the equation $1^x = 2$, it is no longer possible. Whether it is possible or not does not depend on us, but on objective laws, and any alleged creative force is restricted by these.

However, on considering such a case further, just by itself, the argument against creation changes into one for it. Namely, on asking whether the number domain can be extended or not, the question assumes that the number concept is uniquely determined. And then it would seem as if we had to answer the question negatively, as if we arrived somewhere, somehow, at the boundary of the number realm; for the equation $1^x = 2$ just doesn't admit a solution. In reality this is the situation: If a calculus were to be formed by which that equation can be solved, it could be done in a pinch; only this would be a very odd calculus, basically distinct from anything thus far called "number calculus" or "arithmetic." In such a calculus certain fundamental laws of our arithmetic would no longer be valid; for example, a number of this calculus would not become greater through the addition of an ordinary number, etc. — but all this would, after all, not be an objection to it. Of course, it would be rather completely isolated, as if it were a foreign body among the other calculi. For this reason we do not think of such a system as a continuation of our number realm. We express this somewhat vaguely by stating that the number domain cannot be extended in this direction, that there are no such numbers. However, this certainly only means that we forego calling such a calculus a number calculus.

For Frege this was the alternative: either we are dealing with strokes of the pen on paper — this would not give an arithmetic — or we must grant that the symbols have a meaning, and then the meaning exists independently of the symbols. However, the meaning is indeed not an entity which is united in a mysterious way to the symbols; instead, it is the application of the symbols, and *we* have command over these.

Epilogue

Finally, the author gladly fulfils the duty of pointing out those sources—as far as they have not already been expressly mentioned in the text—which he has used in the composition of this book. Above all, these were the lectures on *Elementarmathematik vom höheren Standpunkte aus (Elementary Mathematics from a Higher Standpoint)* by Felix Klein and *Theorie und Anwendung der unendlichen Reihen (Theory and Application of Infinite Series)* by Konrad Knopp.

With regard to the foundations of mathematics, the author has drawn a number of ideas from an unpublished manuscript of Ludwig Wittgenstein which he has been allowed to peruse. I have taken from this work the exposition regarding induction (p. 93 to 99); the criticism of Frege's and Russell's definition of the concept "numerically equivalent" (p. 107 to 113); the ideas presented on p. 119 ff.; furthermore, the remark that the concept of equality does not necessarily have to be transitive (p. 63); the attitude toward certain statements of Brouwer (p. 214 f.); finally, the ideas brought forth (p. 211 f. and p. 224) about the intuitable continuum, the exposition about number as a family of concepts on p. 237, and the criticism of the first argument on p. 238 f. The author wishes to add, however, that he is not entirely sure, chiefly due to the brevity of the presentation, how far his expositions agree with the ideas of Wittgenstein, and on that account he assumes responsibility for his presentation.

INDEX

A

Abridged multiplication, 67
Abscissa, 23
Absolute value, 129, 142
Abstraction, 235
Absurd numbers, 44
Accumulation, point of, 123 ff., 147, 195, 206 ff.
Actual
    classes, 116
    numbers, 103
Actual-infinitesimal, 210, 222 f., 232, 242
Addition, 2, 3, 10, 34, 37, 57 f., 79 ff., 106, 143, 179, 186, 235, 243
    as a combination of sets, 68
    associative, 57, 106
    associative law of, 34, 66 f., 81 ff., 95 f., 98, 106
    commutative, 57, 106
    commutative law of, 34, 66, 82 f., 106
    laws for, 66 f.
    modulus of, 60
    of sequences, 141
    of waves, 62
    recursive definition of, 80, 99
    rules of, 121
    tables, 67
Affine
    geometry, 177, 179, 236
    group, 179
Akkadian, 51
Algebra, 49 f., 95, 102, 226
Algebraic
    equation, 236
    numbers, 16, 212
Algorithms, 9

All, 97
Analysis, 23, 102, 191, 196, 205, 226, 228
Analytic(al), 90
    demonstration, 89
    function, 125
    geometry, 23, 214
    judgments, 89 f.
    propositions, 119
Antecedent, 2, 31 f., 194, 200
Anti-Euclidean geometry, 20
Antinomy, 71 f.
Anzahl, 108
Apagogic proof, 169
Appropriate(ness), 64
Approximately, grammar of the word, 225
Approximating polygon, 158, 163
Approximation process, 192, 216
Archimedean axiom, 209 f., 220, 222 ff., 232
    independence of, 224
    system, 221
Area of a circle, 194 f.
Arithmetic, 15 ff., 18 ff., 52, 69, 75 f., 89, 101 ff., 114, 116, 120 f., 138, 217, 237, 240 ff.
    applications of, 14, 239 f., 244
    axioms of, 76, 103
    concepts of, 70, 75, 83
    consistency of, 76, 102 f.,
    construction of, 13, 79 ff., 88
    development of, 50, 120
    domain of, 90
    elementary, 79, 98
    formal, 39, 76

    formal interpretation of, 238 ff.
    foundation of, 16, 24 f., 66 ff., 103, 115, 118, 121
    laws of, 64, 68, 79, 85, 88, 96, 120, 233, 243
    mean, 124, 139
    nature of, 107
    of differences, 64
    of integers, 15, 41, 52 ff.
    of natural numbers, 27, 30, 41, 66 ff., 105
    of number couples, 29, 41
    of quotients, 64
    of rational numbers, 64, 202
    operations of, 143
    ordinary, 29, 32, 34, 38, 191
    philosophy of, 121
    propositions of, 70, 76, 90, 107, 118, 120
    rules of, 2, 42, 59, 122, 209
    structure of, 15, 42, 68, 76, 78, 196
    system of, 191
    truths of, 81, 121
    universal, 16
Arithmetical
    catastrophe, 115
    concepts, 101 f.
    continuum, 211
    deductions, 69
    formulae, 76, 239
    laws, 89
    model, 23
    operations, 2 f., 60 f., 68, 185, 189
    proposition, 89, 101 f.
Arithmetization, 23, 69

247

INDEX

Associative law, 34, 37 f., 66 ff., 81 ff., 85, 95 f., 98, 189, 229
Asymmetric, 32, 56, 188
Atom, 212, 224
Augustine, 116 f.
Axiom(s), 18, 21 ff., 73-77, 90 ff., 104 f., 191, 199, 224 f., 236
    Archimedean, 209 f., 220, 222 ff., 232
    geometric, 18
    independence of, 223
    linear, of congruence, 75
    of arithmetic, 76, 103
    of Cantor-Dedekind, 214
    of Euclid, 18
    of Euclidean geometry, 22, 24
    of infinity, 115 f.
    of logic, 76
    of logical calculus, 101
    of magnitude, 18
    of non-Euclidean geometry, 21
    of Peano, 69, 101, 103 ff., 120
    parallel, 19, 22 f.
    recursively defined class of, 101
    system of, 74 f.
    truth of, 77, 115
Axiomatic(al)
    investigation, 103
    method, 73, 75, 238
    view, 119

B

Babylon(ians), 50 f.
Basic
    laws of arithmetic, 64, 68, 79, 85, 88, 96, 120, 233, 243
    operations of arithmetic, 143
    propositions of arithmetic, 70, 76, 120
    propositions of Peano, 69, 120
    rules of arithmetic, 2, 42, 59, 122, 209

Berkeley, 155
Bernoulli, Daniel, 125
Bessel, 39
Between(ness), 2 f., 11, 47, 73, 75
Boltzmann, 211
Bolyai, Johann, 20, 74
Bolzano, 195 f.
Bound(ed), 160, 171, 195, 203 f.
Brouwer, 98, 102, 172 f., 214 f., 245

C

Calculating
    laws, 66, 233
Calculating rules, 27, 29, 48, 64, 94, 223, 227, 242
    preservation of, 27
Calculus, 28, 39 f., 55, 61, 64, 93, 95, 98 f., 110, 119, 121, 134, 223, 235 ff., 239, 243 f.
    differential, 147 f., 150 f.
    infinitesimal, 28, 128
    number, 243 f.
    of classes, 69, 218
    of differences, 28, 41
    of natural numbers, 99
    of number couples, 53
    of quotients, 52 f., 60
    of rational numbers, 64
    with cuts, 201
    with laws, 218
    with real numbers, 218
    with sequences, 145, 185, 193
Cantor, Georg, 15, 91, 166, 169 ff., 181, 185, 205 ff., 211, 218
    correspondence of, 171
    -Dedekind, axiom of, 214
    theory of, 185 ff., 205 f., 214
Cardano, 9, 44
Cardinal numbers, 1, 115, 235, 237

    domain of, 41
    sequence of, 105
Carnap, 102, 119
Cartesian system of coordinates, rectangular, 10
Cauchy, 126
Center of
    condensation, 132
    projection, 45
Central projection, 45, 176, 179
Chess, 76, 119, 191, 240 f.
Chuquet, 44
Circle, 19, 160, 177, 194 f., 221 ff.
Class(es), 108 ff., 114, 200-206, 218
    calculus of, 69, 218
    lower (upper), 200-205
    of classes, 108, 114
    of proper fractions, 5
    of rational numbers, 218
Clifford, 211
Closed system, 2 f., 60, 184, 198, 207, 232
Color, 212 f.
    composition, 225
    continuum, 212
Combination of sets, 68
Commutative law, 34, 37 f., 62, 66, 68, 82 f., 86, 99, 189, 233
Complete induction, 88, 91 f.
    inferences by, 82
    principle of, 68, 79 ff., 88 ff.
Completeness, 103
    theorem, 209, 211
Complex number(s), 10, 226 ff., 235
    as number couples, 65, 228 ff.
    geometrical representation of, 10, 227 f.
    higher, 15, 226, 233
    ordinary, 226, 233
    system of, 230, 233
Complex units, 229
Composition, 68, 178 f.
    color, 225

INDEX

of rotations, 62
Computing, practice of, 67
Condensation, center of, 132
Configurations, 22, 77, 175 ff., 180, 224
Congruent, 73, 76, 224
Consequent, 2, 4, 31 f., 194, 200
Consistency
  definition of, 77
  of arithmetic, 102 f.
  of arithmetic and logic, 73, 76
  of classical mathematics, 102
  of Euclidean geometry, 23
  of geometry, 224
  of logico-mathematical systems, 101
  of new logic, 73
  of system of rational numbers, 59
  of theory of numbers, 100
  of theory of real numbers, 217 f.
  problem of, 20, 22, 100, 102
  proof of, 20, 22, 41, 77, 101
Constructibility, 16
Construction, 192
  geometric, 16
  of arithmetic, 13, 79 ff., 88
  of geometry, 24
  of rational numbers, 52, 65
  of real numbers, 24, 101, 145, 205
  of theory of integers, 25 ff., 52
Constructive proof, 97
Continuity, 149, 198 f., 206, 208, 211 f.
Continuous, 173, 199
  correspondence, 160 f., 170-173
  curve, 155-158, 163 f.
  cut, 201
  function, 159 f.
  image of unit segment, 160, 173

motion, 159 f.
one-to-one transformations, 174, 180 f.
set, 206 f.
space, 212
straight line, 198 f.
system, 200 f., 203
Continuum, 213
  arithmetical, 211
  color, 212
  intuitable, 213, 245
  linear, 206, 208
  real, 203, 211, 220
Contradiction, 21, 236
  law of, 169
  principle of, 89
Converge to, 128, 131, 141 f., 144 f., 151, 190, 215, 219
Convergence, 145, 183
Convergent processes, 162, 192
Convergent sequence(s), 135, 144 ff., 183 ff., 187 ff., 192 f., 196 f., 210
  as real numbers, 189, 196
  definition of, 128, 184
  domain of, 143
  examples of, 131
  sum of, 141 f., 189
Coordinate(s), 149, 155, 171, 173
  geometry, 23, 148
  plane, 10
  system of, 10, 212
Correspondence
  continuous, 160, 170-173
  discontinuous, 171
  isomorphic, 43, 60, 190, 202, 205, 209 ff.
  many-valued, 171
  of Cantor, 171
  of Peano, 171
  one-to-one, 42, 45, 60, 169, 171 f., 190, 202, 205, 209 ff., 214
  similar, 43, 60, 190, 202, 205, 209 ff.
  single-valued, 43, 171
Counting, 1, 44, 112 f., 118, 213
  numbers, 41, 117

Cube, 180
Curvature, 223
Curve(s), 153 ff.
  continuous, 155-158, 163 f.
  definition of, 160, 173 f.
  differentiable, 157 ff., 164
  intuitive, 165
  Jordan, 159 f.
  of H. v. Koch, 158 f.
  of Peano, 160-164
Curvilinear angle, 221
Cut, 201-206, 210 f.
Cyclic order, 46

D

d'Alembert, 150
Decidable, 102
Decimal fraction(s), 214, 216 f.
  development, 194
  infinite, 98, 194, 217
  molecules of, 169
  non-periodic, 8
  non-terminating, 8, 93, 167 f.
  periodic, 8
  representation of numbers as, 8, 167
  terminating, 8, 168 f., 171
Decomposition of unit fractions, 50
Dedekind, 90, 170, 185, 198 f., 201, 204 ff., 210, 212 f.
  axiom of Cantor-, 214
  cut, 205, 210, 218
  theory, 198 ff., 206, 214
Deductive system(s), 119, 166
Definition
  demonstrative, 165
  formation of, 80, 85, 99
  inductive, 91
  recursive, 80, 85
  requirement of, 165
Dense, 4, 116, 207 f.
  system, 4, 8, 199
Derived set, 207

249

INDEX

Descartes, 23, 148
Descriptive science, 71
Determination, 235
Development of
    arithmetic, 50, 120
    logic, 72
    logical calculus, 76
Difference(s), 27 f., 55, 87, 145, 184, 232
    arithmetic of, 64
    calculus of, 28, 41
    difference of, 27
    number couple as, 37, 42
    of number couples, 34, 36, 228
    of quotients, 54
    positive numbers as, 28, 42
    product of, 27
    quotient, 150, 157
    sequence, 144, 185, 188, 197
    sum of, 27
Differential calculus, 147 f., 150 f.
Differential quotient, 141 ff., 157
    left-hand, 153
    right-hand, 153
Dimension, 161, 166, 173 f., 181, 224, 233
    invariance of, 172
    number, 173 f.
Diophantine equation, 40
Directed quantity, 226, 234
Direction(s), 80 f., 85, 99, 116, 119, 153, 156, 163 f., 171, 180, 223, 225, 238, 244
Discontinuous, 156, 171, 201, 211 f.
Disjunction, 1, 56, 84, 186 f., 189, 220, 225
Distortions, 174 f., 180
Distributive law, 58, 67, 85, 106, 230
Divergent
    sequence, definition of, 128
    series, summing of, 140

Divisibility, 7, 86, 106, 126
Division, 2 f., 12, 40, 52-55, 58 f., 86 f., 92-94, 143, 145, 147, 161 f., 195, 199, 204, 229, 231, 233
    by zero, 56, 59, 143
    without remainder, 2, 52 f.
Dodecagon, 194
Domain of
    analysis, 23
    arithmetic, 90
    cardinal numbers, 41
    complex numbers, 11, 13, 226
    integers, 40, 44, 52 f., 61
    logical structures, 71
    natural numbers, 2, 61
    number couples, 29 f., 34, 36, 57
    rational numbers, 189, 203
    real numbers, 13, 196 f., 203, 208
    sequences, 131
    symbolism, 77
Drosophila, 75

E

Egypt(ians), 49 ff.
Element of group, 178, 236
Ellipse, 147, 177
Empirical
    cognition, 110
    proposition, 118
Empty
    set, 174
    thing, 74
English mathematicians, 44, 67
Enlargement, 176
Equal (to), 1, 18, 29 ff., 33, 53-57, 59 f., 84, 107, 142, 145, 170, 185, 188, 190, 197 f., 202 f., 206, 213, 220, 225, 228, 239
Equality, 31, 54, 63, 80, 112, 186, 202, 232, 245
    sign, 107, 119, 238

Equations
    cubic, 9
    Diophantine, 40
    linear, 23
Equilateral triangle, 158, 175, 177
Erlanger Programm, 180
Essential property, 179
Estimation, 225
Euclid('s), 18
    axioms of, 18
    definition of point, 73 f.
    Elements, 212
    structure of, 23
    system, 73, 212
Euclidean geometry, 19 ff., 236
    axioms of, 22, 24
Even numbers, 43, 104, 126, 215
Evolution, 12
Existence, 14 ff., 24, 50, 84, 87, 97 f., 117, 183, 195, 204
    criterion for, 242
    proofs, 97
    statement, 97 f.
Experiment, 110, 121
Extension of
    concept, 27, 48
    convergence, 145
    integers, 60
    natural numbers, 3, 41
    number concept, 8
    number domain, 5, 12 ff., 45, 47 f., 185, 201, 226, 243
    numbers, 12 ff.
    principal group, 180
    rational number system, 209
    real number system, 209
    sum, 82
Extension(s), 110, 217, 233
Extensive quantity, 224
Euler, 9, 125
    relation of, 226

F

Family, 237, 245
    similarity, 237

INDEX

ermat, 23, 130 f., 214 ff.
   problem of, 130 f., 215 f.
ctitious numbers, 9, 44
eld, 4, 143
nite, 94 ff., 178, 196
   computation, 94
   disjunction, 86
   induction, 78
   intuitive methods, 100
   sequence, 78
   series, 139
   sets, 170, 181
   sums, properties of, 139
rces, 62
rmal
   arithmetic, 39, 76
   interpretation of arithmetic, 238 f.
   logic, 89
   system, 101 f.
malism, 39, 51, 100 ff., 07, 116, 223
malist, 97, 104
malistic-lateral framework of analysis, 191
ndations, 100 ff.
   for geometry, 103, 223
   of arithmetic, 16, 24 f., 66 ff., 103, 115, 118, 121
   of mathematics, 100, 120
r-color problem, 181
-dimensional
   space, 74, 180
   space-time world of Minkowski, 234
-term numbers, 234
tional numbers, 3, 49, 57, 220
   existence of, 14
   set of, 116
ions, 4, 6, 49, 52, 56, 136 ff., 151
   decimal, see decimal
   decomposition of unit, 50
   atural, 50
   atural ordering of, 138
   egative, 138

operational rules of, 52
   proper, 5, 203
Frege, 14, 69 ff., 76, 88 ff., 107 f., 111 f., 114 ff., 119, 121, 238-242, 244 f.
French mathematicians, 67
Function(s), 151, 154 ff., 159, 160, 219 f., 227
   continuous, 159 f.
   degree of, 220
   expansion of, 125
   infinitely many-valued, 227
   measure of, 220
   sequence of, 220
   sine, 154 ff.
   single-valued, 160
   theory of, 102, 227

G

Game, 77, 239 f., 242
   arithmetical, 240
   of formulae, 239
   of proofs, 77
   with symbols, 238
Gap(s), 201 f.
Gauss, 10, 39, 228, 233 f.
Gaussian interpretation, 10
Gentzen, 102
Geography, 175, 181
Geometric(al)
   axioms, 18
   construction, 16
   elements, 74
   property, 175 ff., 212
   propositions, elementary, 75
   relations, 23
   representation of complex numbers, 10, 227 f.
   representation of imaginary numbers, 10
   series, 123
   system of axioms, 74
Geometry, 16, 73, 75, 89 f., 105, 175 ff., 212, 214, 224, 226, 237
   affine, 177, 179
   analytical, 23, 214

anti-Euclidean, 20
   construction of, 24
   coordinate, 23, 148
   Euclidean, 19 ff., 236
   foundation for, 103, 223
   metric, 177, 179 ff., 236
   modern, 74
   Archimedean, 223 f.
   non-Euclidean, 20-23, 73 f.
   of inversions, 180
   of one-to-one and continuous point transformations, 180
   of one-to-one point transformations, 181
   ordinary, 177, 179
   projective, 177, 179, 236
   structure of, 17 f.
Germany, 67
Gleichzahlig, 107
Gödel, 100 f., 217 f.
Goldbach, 125
Grammar, 118, 225
Grandi, Guido, 123 ff., 140
Grassmann, H., 68, 233
Greater (than), 1, 4, 11, 32 ff., 45 f., 53-56, 60, 62, 84, 145, 170, 185, 188, 190, 194, 202, 206, 219 ff., 225, 228, 230 f., 243
Greatest lower bound, 203 f.
Greeks, 50, 117, 147
Group, 177-180
   affine, 179
   element of, 178, 236
   principal, 179 f.
   projective, 179
   transformation, 180

H

Hahn, 119
Hamilton, 67 f., 119, 228
Hankel, H., 16, 27, 39, 45, 67, 191
Helmholtz, 173

251

INDEX

Heredity, laws of, 75
Hexagon, 194
Higher complex numbers, 15, 226, 233
Hilbert, 18, 23, 75 ff., 100, 102, 160, 224
Hindus, 44
History of mathematics, 9
Horn-shaped angles, 221 f
Hyperbola, 147, 177
Hypercomplex numbers, 226 ff.
Hypothetical-deductive system, 74

I

Icositetragon, 194
Ideal
    lines, 73
    planes, 73
    points, 73
Ideography, 51
Image(s), 1, 10, 21, 46, 145, 160, 163 f., 169, 171, 173
Imaginary numbers, 8 ff., 12, 226, 228, 231, 241
    geometrical representation of, 10
Immanent criterion, 184
Impossible numbers, 9
Improper
    limiting value, 145
    point, 46 f.
Incident, 73
Incidental property, 179
Increment, 149
Independence of axioms, 103, 223 f.
Indeterminacy, zones of, 225
Induction, 93, 98, 229, 245
    definition by, 91
    finite, 78
    mathematical, 91 f.
    method of, 88, 213
    principle of complete, 68, 79 f., 82, 88 ff.
    proof by, 91 f., 95-99
Inertial paths, 103
Infinitarily, 219 f.

Infinite, 5, 84, 86, 91, 93-96, 99, 102, 124 ff., 133, 167, 170, 178, 196, 219 f.
    actual, 126
    cardinal numbers, 91
    decimal fractions, 167, 216
    extension, 94
    on both sides, 3
    on one side, 2
    potential, 126 f.
    processes, 123
    sequences, 102, 141
    series, 123 ff., 138 ff., 180
    sets, 170, 181, 195, 203, 217
    smallness, order of, 146
Infinitely
    distant point, 46
    small quantity, 150
Infinitesimal calculus, 28, 128
Infinity, 22, 88 f., 93, 96, 98 f., 124-128, 131, 157, 170, 213, 216, 220
    axiom of, 115 f.
    of syllogisms, 88 f., 91
    order of, 219
Inflexive, 84
Innumerable, 213
Insertion, process of, 4, 6, 189, 208 f.
Integer(s), 3 f., 6 f., 25 ff., 56-61, 190, 208, 235, 237
    arithmetic of, 15, 41, 52 ff.
    as number couples, 28 ff., 65
    as rational numbers, 60
    calculus of, 64
    characterization of, 106
    complement of, 50
    construction of, 25 ff., 52, 65
    divide a number couple by, 229
    domain of, 40, 44, 52 f., 61
    existence of, 14

    extension of, 60
    laws of, 120
    multiply a number couple by, 58, 229, 232
    reciprocal of, 50
    series of, 210
    system of, 3, 27, 42 f., 60
    theory of, 25 ff.
Intuitable continuum, 213, 245
Intuition, 18 f., 23 f., 68, 73, 78, 153, 159, 164, 166, 212, 214, 221, 225
Intuitionist, 97, 196
Intuitive
    conclusions, 76
    continuum, 224
    meaning of addition, 68
Invariance of dimensions, proposition of, 172
Invariant, 175, 179 ff.
Inverse
    element, 179
    number couples, 35, 37
Inverse operation(s) of, 64, 67
    addition, 58
    multiplication, 40, 58, 231
Inversion transformation, 180
Inversions, geometry of, 180
Irrational limiting value, 184
Irrational numbers, 7 f., 10, 12, 185, 192, 197, 200, 208, 214, 216, 242
    as a rule, 192
    existence of, 14, 16, 24, 182
    formal theory, 191
    system of, 8
Irrational points, 138
Irrational sequence, 197
Irreflexive, 32, 56, 188
Isolated points, 206, 208
Isomorphic systems, 65, 74
Isomorphism, 43

252

INDEX

J

Jordan, C., 159 f.
Jump, 201
Jürgens, 172
Justification, 15, 64, 212
Justified, 30, 34, 64, 67, 205, 213

K

Kant, 68, 89 f., 119
-dimensional continuum, 172
Klein, F., 20, 90, 159, 175, 180, 245
Knopp, K., 245
Knots, 180
Koch, H. v., 158
König, 78, 169
Kowalewski, G., 134

L

Lambert, 20
Language, 49, 51 f., 68 f., 114 f., 117, 119, 127, 165 f., 174, 213, 237, 239
Law(s), 94, 129 ff., 135, 147, 149, 151, 190, 192, 194, 216 ff.
  arithmetical, 89
  associative, see associative law
  calculating, 66, 233
  calculus with, 218
  commutative, see commutative law
  distributive, 58, 67, 85, 106, 230
  of monotony, 34, 39, 57 f., 62, 66 f.
  of addition, 66 f.
  of arithmetic, 64, 68, 79, 85, 88, 96, 120, 233, 243
  of contradiction, 169
  of differential calculus, 151
  of excluded middle, 71, 103
  of heredity, 75
  of integers, 120
  of logic, 76, 89, 121
  of multiplication, 66 f.
  of natural numbers, 120
  of rational numbers, 120
  of real numbers, 120
Least upper bound, 203 f.
Leibniz, 9, 124 f., 140, 148, 150
Lexicographic order, 106
Limit, 67, 121, 123 ff., 142 ff., 149, 151 f., 161, 164, 192, 195, 204, 217, 219
  definition of, 134
  operation, 143, 184, 196, 198, 206, 208, 210, 217
  point, 131, 133
Limiting position, 148, 156 f., 159, 164
Limiting value, 125-129, 132-135, 138-147, 150 f., 183 f., 190, 192, 196
  definition of, 128, 134
Linear
  combination of units, 229 f., 232
  congruence axioms, 75
  continua, 206, 208
  equation, 23
  magnitude, 191
  order, 11, 47, 62, 106
  path, 163
Linearity, 173
Lobachevski, 20
Logarithm, 227
Logic, 69-73, 89, 91, 100, 102, 118 f., 121, 224, 231, 235
  axioms of, 76
  consistency of, 73, 76
  formal, 89
  laws of, 76, 89, 121
  propositions of, 70 f.
  rules of, 68, 102
  symbolic, 236
Logical, 18, 20 f., 24, 68-76, 78 f., 90, 95 f., 103, 107, 114, 118, 120 f., 182, 192, 224
  definition of number concept, 121
  grammar, 119
  school, 107 ff., 121

Logical calculus, 76
  axioms of, 101
  theory of, 101
Logical-formal, 75
Logicalization, 69, 71
Logician, 88, 118
Logico-mathematical systems, 101
Loi de continuite, 148
Lorentz transformation, 234
Lower class, 200-205

M

Magnitude, 45, 191 f.
  axioms of, 18
  linear, 191
Manifold, 74, 180, 233
  two-dimensional, 11
Many-valued function, 227
Mathematical, 24, 39, 50 f., 68 f., 73, 100 f., 107, 116, 119, 124, 127, 157, 204, 234
  concepts, 51
  continuum, 212, 224
  formalism, 51
  induction, 91 f.
Mathematician, 9, 14 f., 17, 19 f., 23 ff., 41, 44, 46, 49, 50, 64, 67 f., 73, 88, 90, 100, 123, 125, 129, 150 f., 165 f., 177 f., 190, 198, 213 f., 236, 241 f.
Mathematics, 9, 27, 30, 50 ff., 69, 70 f., 73, 76 f., 98, 100, 102, 111, 116, 118 ff., 123, 128, 147 f., 159, 165, 191, 196, 215, 225-228, 236, 240, 242, 245
  one-dimensional, 172
  n-dimensional, 172
Mechanics, 71
Menger, 74, 165, 173
Mensuration of angles, 225
Metamathematics, 77, 100
Metric geometry, 177, 179, 180 f., 236
Mill, 118
Minkowski, 234

253

INDEX

Minor, 88
Minorant, 203 f.
Mirror reflections, 176
Möbius strip, 180
Model, 4, 18, 21, 23, 41, 45, 74, 185
Modulus, 60
Molecules, 169
Monotone sequences, 193
Monotonic
   decreasing sequence, 193 f.
   increasing sequence, 193 f.
Motion, continuous, 159 f.
Motions in space, 179
Multiple of a number couple, 36, 38, 58, 229, 232
Multiplication, 2 f., 38 ff., 59, 79, 85, 106, 143, 178, 229 f., 232 f.
   abridged, 67
   associative, 106
   associative law of, 38, 66 f., 85, 106
   commutative, 106
   commutative law of, 38, 66, 86, 99, 106, 233
   laws for, 66 f.
   modulus of, 60
   monotony law of, 48, 58, 66
   rules, 231
   tables, 67
Musical pitch, 63

N

Natural
   fractions, 50
   ordering, 42, 138
   science, 103
Natural numbers, 1-4, 12, 34-37, 40-43, 48, 52, 59, 66 f., 70, 79, 87, 91 f., 98, 106, 115, 130, 209, 218, 235
   arithmetic of, 27, 30, 41, 66 ff., 105
   differences of, 28
   domain of, 61
   equality of, 29 f., 63
   extension of, 3, 41

   geometric representation of, 1
   laws for addition of, 66 f.
   laws for multiplication of, 66 f.
   laws of, 120
   multiplication of, 39
   pairs of, 28 f.
   sequence of, 43, 136, 214
   system of, 1-3, 13, 27, 29 f., 40, 42 f., 60, 66
   theory of, 101, 115
n-dimensional space, 232
Nearly all, 134
Negative
   cut, 202
   fraction, 138
   limiting value, 144
   number couple, 32, 34
   rational number, 186, 200
   sequence, 144, 186-189
   unit, 36
Negative numbers, 3, 9, 12-15, 28, 41-46, 49, 65, 242
   as differences, 28
   as supra-infinite, 44, 46
   existence of, 14 f.
   generation of, 3
   notation for, 49
   representation of, 3, 14 f., 45 f.
Nested
   intervals, 162, 195
   squares, 162
Nests of intervals, 163, 182 f., 193 f., 205, 210
Neugebauer, 52
Newton, 148, 150
Non-Archimedean
   geometry, 223 f.
   system, 224
Non-closure
   of concepts, 237
   of mathematics, 102
Non-constructive proof, 97
Non-differentiable, 159

Non-Euclidean geometry, 20-23, 73 f.
Non-normal set, 72
Normal
   classes, 72
   form of number couples, 36, 229
   set, 72
n-th multiple, 35
n-tuples of numbers, 232
Null
   class, 115
   couple, 32, 34, 37
   cut, 202
   sequence, 144-147, 149, 151 f., 157, 185-189, 197 f.
Number, 103 f., 107 f., 114-118, 121, 235 f.
   as a law or rule, 190, 192, 216
   calculus, 243 f.
   continuum, 212
   extension, 8
   types, 235, 237
Number axis, 5-8, 10, 16, 45 ff., 132, 145, 190, 192, 203, 205, 210, 213
Number circle, 45 ff.
Number concept
   extension of, 8
   logical definition of, 121
Number couple(s), 28-42, 55-60, 64, 186, 228-232
   arithmetic of, 29, 41
   as difference of positive numbers, 37, 42
   as linear combination of units, 229
   calculus of, 53
   components of, 228
   difference of, 34, 36, 228
   division of, 40, 59, 231
   domain of, 30, 34, 37, 57
   equality of, 29 ff., 57, 64, 228
   forms of, 32 ff., 36, 56 f.
   inverse, 35, 37

meaning of, 65
multiple of, 36, 38, 58, 229, 232
multiplication of, 38, 58
negative, 32, 34
normal form of, 36, 229
operational rules of, 41, 53
opposite, 35
positive, 32, 34
product of, 34, 38, 230
subtraction of, 37
sum of, 34 f., 37 f., 57 f., 228
system of, 30
number domain
closure of, 3
extension of, 5, 12 ff., 45, 47 f., 185, 201, 226, 243
sequence, 127, 213
system, 2, 47 f., 205, 209 f., 219, 223, 232 f.
triples, 64
numbers
absurd, 44
actual, 103
algebraic, 16, 212
cardinal, 1, 115, 235, 237
complex, 10, 226 ff., 235
counting, 41, 117
divisibility properties of, 7
even, 43, 104, 126, 216
extension of, 12 ff.
fictitious, 9, 44
four-term, 234
hypercomplex, 226 ff.
images of, 1, 10, 145
imaginary, 8 ff., 12, 226, 228, 231, 241
impossible, 9
irrational, see irrational numbers
natural, see natural numbers
negative, 3, 9, 12-15, 28, 41-46, 49, 65, 242

negative rational, 186, 200
n-tuples of, 232
odd, 126, 215
positive, 3, 9, 15, 41, 45 f., 65, 209, 215, 218
positive imaginary, 10
positive rational, 137 f., 186, 200
positive real, 10
rational, see rational numbers
real, see real numbers
relatively prime, 6 f.
theory of, 100
transcendental, 16
types of, 1 ff., 25
Number series, 1 f., 68, 84, 86, 99, 104 f., 113, 120 f.
formation rule of, 3
Numeral(s), 41, 92 f., 95, 99, 103, 113 f., 117 ff., 161 f., 167 f., 216 f.
system of, 44
Numerically equivalent, 107-112, 114, 245

O

Occident, 44
Occidental mathematics, 226
Ohm, M., 45
One-dimensional, 166, 173 f.
One-sided surface, 181
One-to-one
correspondence, see correspondence
isomorphism, 230
point transformation, 180
relation, 108, 111 f.
Operating
rules, 193
with letters, 44
with sequences, 141 ff., 189, 218
Operation(s), 38, 51, 61, 79 f., 102, 143, 178, 237

arithmetical, 23, 60 f., 68, 185, 189
limit, see limit
of arithmetic, 143
Operational rules of
fractions, 52
number couples, 41, 53
rational numbers, 60
Opposite number couples, 35
Order, 47, 62, 65, 68
cyclic, 46
lexicographic, 106
linear, 11, 47, 67, 106
of infinite smallness, 146
of infinity, 219
of poles, 220
Ordered, 4, 27, 59, 140, 200, 209
system, 1, 3 f., 33, 56, 188, 201, 221
Ordering, 60, 78, 137, 200
natural, 42, 138
of real numbers, 46 ff.
Ordinal numbers, system of, 221
Ordinary
arithmetic, 29, 32, 34, 38, 191
complex numbers, 226, 233
geometry, 177, 179
multiplication, 38
numbers, 10
Ordinate, 23
Origin, 155 ff., 219
Oscillating sequence, 193
Oscillations, 156
Osculating circles, 223

P

Pairs of
natural numbers, 28 f.
real numbers, 23
sequences, 193
Parabola, 147, 177
Parallel(s), 22, 73, 177
axiom, 19 f., 22
projection, 177, 179
Partial sums, 138 f.
Pasch, 18

255

INDEX

Peano('s), 68 f., 76, 103 ff., 120, 160, 163 f., 166, 171
   axioms, 69, 103 ff., 120
   correspondence, 171
   curve, 160-164
Pentagon, 112
Perfect set, 207 f.
Periodic division, 93 f.
Periodicity, 93 ff.
Permanence
   of properties of equality, 64
   principle of, 27 f., 45, 140
Philosophy of arithmetic, 121
Phonetic writing, 51
Photography, 176 f.
Physical space, 212
Physics, 224 ff.
Pictorial writing, 51
Pieri, 74
Plato, 71, 228
Poincaré, 72, 88, 90 f., 98, 119, 173
Point
   definition of, 73 f.
   ideal, 73
   image, 145, 164, 169, 171
   improper, 46 f.
   irrational, 138
   isolated, 206, 208
   meaning of, 236
   of accumulation, 123 ff., 147, 195, 206 ff.
   of infinity, 219
   rational, 7
   representation as, 1, 3, 10
   representation of, 23
   series, 1 f.
   set, 5, 195, 206 f.
   transformation, 180 f.
Pole, 219 f., 227
   order of, 220
Polygonal areas, 194 f.
Polynomials, 106
Porous, 207
Positive
   cut, 202

imaginary number axis, 10
limiting value, 144
number couple, 32, 34
rational numbers, 137 f., 186, 200
real number axis, 10
sequence, 144, 186-189, 206
units, 36
Positive numbers, 3, 9, 15, 41, 45 f., 65, 209, 215, 218
   as differences, 28, 42
   as number couples, 42
   axis of, 144
   natural numbers corresponding to, 42 f.
   points on number circle corresponding to, 45 f.
Postulational method, 14
Potential infinite, 127
Power, 25 f.
   rules, 26
Preciseness, 165 f.
Predecessor, 2, 4, 84
Primary laws, 68
Prime number, 126, 130
Primitive system, 78
Principal group, 179 f.
   extension of, 180
Principle of
   complete induction, 68, 70 f. 82, 88 ff.
   contradiction, 89
   permanence, 27 f., 47 f., 229, 233
Probability, 121
Problem, 237
   Fermat's, 215
   four-color, 181
   of consistency, 20, 22, 100, 102
   of decomposing a number, 44
   of solving equations, 44
   of tangents, 147
Process
   infinite, 123
   of becoming infinite, 219 f.
   of insertion, 4, 6, 208 f.

Product, 27, 34, 38 f., 43, 55, 230, 232 f., 236
Progression, 104
Projection
   center of, 45
   central, 45, 176, 179
Projective
   geometry, 177, 179, 236
   group, 179
Proof(s), 25, 91, 95, 99, 237
   apagogic, 169
   by induction, 91 f., 95-99
   constructive, 97
   existence, 97
   game of, 77
   indirect, 19
   method of, 91
   non-constructive, 97
   of consistency, 20, 22, 41, 77, 101
   of transitivity, 55
   theory of, 77
Proper fractions, 5, 203
Property, 218
   arithmetical, 89, 101 f
   essential, 179
   geometrical, 175 ff., 212
   incidental, 179
Proportion, 53
Propositions
   empirical, 118
   geometric, 75
   of arithmetic, 70, 76 90, 107, 118, 120
   of invariance, 172
   of logic, 70 f.
   of Peano, 69, 120
Pseudo-problem, 125
Psychological, 93
Pyramid, 180
Pythagoras, 5, 15, 214
   theorem of, 6

Q

Quadratic equation, 46
Quaternions, 233 f., 242
Quotient(s), 3, 53 ff., 6 87, 92, 94, 146 f., 149
   arithmetic of, 64
   calculus of, 52 f., 6

difference of, 54
differential, 141 ff., 157
of infinitely small quantities, 150

adians, 154
amsey, 119
atio, 149 f., 219, 222
ational
  cut, 203
  limiting value, 183 f., 190
  points, 5, 7, 182
  real number, 190, 202 f.
tional number 0, 60
tional number 1, 60
tional numbers, 4-8, 10, 49 ff., 184, 186, 189 f., 203 f., 208-211, 215-218, 229, 235
  arithmetic of, 64, 202
  calculus of, 64
  classes of, 218
  consistency of system of, 59
  construction of, 52, 65
  domain of, 189, 203
  formal theory of, 191
  generation of, 192, 217
  incompleteness of, 210
  laws of, 120
  negative, 186, 200
  operational rules of, 60
  positive, 137 f., 186, 200
  representation of, 4
  separation of, 200
  sequences of, 137 f., 185, 196 ff.
  structure of, 5, 211
  system of, 3 f., 8, 50, 53, 59 f., 65, 184, 196, 199-202, 209, 211, 217
  visualization of, 213

ontinuum, 203, 211, 220

cut, 203
unit, 234
Real number(s), 182 ff., 219, 223-232, 235
  as a rule (law), 216 f.
  as cuts, 203, 205 f., 210
  as decimal fractions, 8, 167, 216 f.
  as points, 10, 214
  as sequences, 185, 189, 191, 197, 206, 215
  calculus with, 218
  complex number as a couple of, 228
  consistency of, 217 f.
  construction of, 24, 101, 145, 205
  degrees of, 196
  domain of, 13, 196 f., 203, 208
  extension of, 209
  geometrical model of, 45 f.
  laws of, 120
  ordering(s), of, 46 ff.
  points as pairs (triplets) of, 23
  rational, 190, 202 f.
  scale of, 220
  sequence of, 196 f.
  system of, 8, 198, 203, 208-211, 217, 230
  theory of, 23 f., 217 f.
Realization of system of axioms, 75
Reciprocal, 50, 59, 150, 221
Rectangular Cartesian system of coordinates, 10
Recurrence, reasoning by, 89 ff.
Recursive
  definition, 80, 84 ff., 99, 101
  formula, 131
  process, 174
Reduction
  process of, 218
  transformation, 176

Reflection(s), 176, 179 f.
Reflexive, 30, 55, 63, 186
Regular solids, 105
Reiff, 124
Relation(s)
  Euler's, 226
  geometric, 23
Relatively prime numbers, 6 f., 106
Relativity, theory of, 234
Remainder, 92 ff.
Renaissance, 44, 214
Representation
  as decimal fractions, 8, 167
  as points, 1, 3, 10
  of complex numbers, 227 f.
  of imaginary numbers, 10
  of natural numbers, 1
  of negative numbers, 3, 14 f.
  of points, 23
  of rational numbers, 4
  on number circle, 145
Rest, 179
Resultant of forces, 62
Reymond, P. du Bois, 182, 191
Riemann, 173
Rotary extension, 234
Rotation(s), 61 f., 178, 234
Rubber surface, 174
Rudimentary number concept, 78
Rule(s), 119, 130 f., 190, 192, 216 f., 238
  calculating, 27, 29, 48, 64, 94, 223, 227, 242
  of addition, 121
  of calculus, 239
  of logic, 68, 102
  of rational numbers, 60
  of signs, 15, 39, 44
  power, 26
Russell, 14, 71 f., 76, 91, 104, 114 ff., 119, 205

INDEX

S

Scale of
  points, 1
  real numbers, 220
Schlick, 63, 74
Schopenhauer, 169
Schubert, H., 107
Science, descriptive, 71
Secant, 148, 165 f., 159, 164
Segment, 205, 210
Sentence, 236
Sentential content, 70
Separation, 167 f., 170, 191, 200 f., 203, 210 f., 235
Sequence(s)
  addition of, 141
  calculus with, 145, 185, 193
  difference, 144, 185, 188, 197
  divergent, 128
  infinite, 141
  irrational, 97
  monotonic decreasing, 193 f.
  monotonic increasing, 193 f.
  negative, 144, 186-199
  null, see null
  of cardinal numbers, 105
  of discourses, infinite, 102
  of even numbers, 43
  of formulae, finite, 78
  of functions, 220
  of natural numbers, 43, 136, 214
  of numbers, 128-131, 183, 186, 218
  of positive rational numbers, 137 f.
  of rational numbers, 137 f., 185, 196 ff.
  of real numbers, 196 f.
  of quotients, 146, 150
  of sequences, 197
  operating with, 141 ff., 189, 218
  oscillating, 193
  positive, 144, 186-189, 206
  real numbers as, 185, 189, 191, 206, 215
Series
  divergent, 140
  finite, 139
  of equations, 99
  of geometries, 181
  of inferences, 91
  of integers, 210
  of proofs, 98
  of rules, 80
Servois, 67
Set(s), 72
  combination of, 68
  continuous, 206 f.
  derived, 207
  empty, 174
  finite, 170, 181
  infinite, 170, 181, 185, 203, 217
  number of a, 108
  of all abstract concepts, 72
  of fractional numbers, 116
  perfect, 207 f.
  theory of, 8, 69, 73, 181, 207
Signs, rule of, 15, 39, 44
Similar, 19, 176
  correspondence, see correspondence
Similarity, 175, 190, 209, 217, 230, 236
  family, 273
  transformations, 179
Simultaneity, 114
Sine curve, 153 f.
Single-valued
  composition, 178
  correspondence, see correspondence
  function, 160
  image of unit segment, 160, 173
  relation, 108
Skolem, 79, 87, 105 f.
Slope
  average, 149
  instantaneous, 149
Smaller (than), 1, 4, 11, 18, 33, 46, 53-56, 60, 62, 83 f., 142, 145, 170, 185, 188, 202, 206, 220 f., 225, 228, 230 f.
Solvable, 102
Space, 163, 177, 206, 211
  continuous, 212
  discontinuous, 211 f.
  Euclidean, 173
  four-dimensional, 74, 180
  n-dimensional, 232
  physical, 212
  three-dimensional, 12 74, 174, 180, 232, 234
  -time of Minkowski, 234
Spatial
  -intuitive, 75
  perceptions, 24
  transformations, 179
Spectrum, 212
Sphere, 62, 177 f., 180
Splitting, 167, 169
State of becoming, 126
Statements, existence, 97
Statistical series, 217
Stifel, M., 44
Stolz, 67
Straight line, 1, 18, 21, 2. 46 f., 74, 104, 149, 15. 169, 176 f., 180, 19. 198 f., 206, 208, 214 221
Structure of
  arithmetic, 15, 42, 6 76, 78, 196
  Euclid, 23
  geometry, 17 f.
  logic and arithmetic 78
  rational numbers, 211
  speech and writing 50
Study, 23
Subject-predicate form, 115
Substitution rule, 98
Subtraction, 2 f., 12, 2 35 ff., 58 f., 61, 86 106, 143, 217
Successor, 69, 78, 80, 8 103 f.

...m, 27, 34-38, 55, 57 f., 61 ff., 66, 68, 80 ff., 121, 123 ff., 138-141, 143, 145, 189, 190, 202, 206, 228, 232, 236
  absolute value of, 142
  partial, 138 f.
  properties of finite, 139
...m of
  colors, 62
  convergent sequences, 141 f., 189
  cuts, 202
  differences, 27
  infinite series, 125, 138 ff.
  notes, 63
  number couples, 34 f., 37 f., 57 f., 228
  rotations of sphere, 61
  sequences, 143
...nerian, 51
...mable, 139 f.
...erimposing, 157
...ra-infinite numbers, ..., 46
...face, 172 ff., 181
...pended classes, 165 f.
...ables, 51
...ogism, 71, 76, 88 f., ..., 98
...bolic
  language, 69, 76
...logic, 236
  writing, 51
...bolism, domain of, 77
...bols
  game with, 238
  of logical calculus, 76
...metric, 30, 55, 63, 186
...hetic a priori judg-...nts, 89 f., 98, 118
...m
...rchimedean, 221
...losed, 2 f., 60, 184, 198, 207, 232
...ontinuous, 200 f., 203
...ecimal, 93
...ense, 4, 8, 199
...ypothetical-deductive, 74
...omorphic, 65, 74

non-Archimedean, 222
number, see number
ordered, 1, 3 f., 33, 56, 188, 201, 221
point, 2
System of
  arithmetic, 191
  axioms, 74 f.
  complex numbers, 230, 233
  concepts of Euclidean geometry, 20
  coordinates, Cartesian, 10, 212
  Euclid, 73, 212
  formulae, 102
  integers, 3, 27, 60, 42 f.
  integers and fractional numbers, 3
  natural numbers, 1 ff., 13, 27, 29 f., 40, 42 f., 60, 66
  number couples, 30
  numerals, 44
  ordinal numbers, 221
  positive and negative numbers, 3
  rational and irrational numbers, 8
  rational numbers, see rational numbers
  real numbers, 8, 198, 203, 208-211, 217, 230
  symbols and operations, 239
  weights and measures, 49

T

Tangent, 147 f., 156, 164, 220 ff.
Tautology, 70 f., 110, 118 f., 236
Tetrahedron, 177 f.
Three-dimensional space, 12, 74, 174, 180, 232, 234
Three-place relation, 86
Theorem
  completeness, 209, 211

  of Bolzano-Weierstrass, 195 f.
  of Pythagoras, 6
  uniqueness, 210
Theory of
  Cantor, 185 ff., 205 f., 214
  forms, 16
  functions, 102, 227
  integers, 25 ff.
  invariants, 180
  irrational numbers, 191
  logical calculus, 101
  natural numbers, 101, 115
  numbers, 100
  probability, 124, 217
  proof, 77
  proportions, 53
  rational numbers, formal, 191
  real numbers, 23 f., 217 f.
  relativity, 234
  sets, 8, 69, 73, 181, 207
  types, 72
Threshold value, 128
Time, 116 f., 159, 163, 171, 211 f.
Topography, 175, 181
Topology, 103, 174, 180 f., 236
Torus, 180
Transcendental numbers, 16
Transfinite
  methods, 103
  numbers, 15
Transformation(s), 82, 176 f., 179 ff.
  inversion, 180
  Lorentz, 234
  of principal group, 180
  similarity, 179
Transitive, 30, 32, 55 f., 63, 84, 186, 188, 245
Transitivity, 55 f.
Two-dimensional, 166, 174, 180
  manifold, 11
Two-sided surface, 181

INDEX

Types
  number, 235, 237
  of numbers, 1 ff., 25
  theory of, 72

U

Ultrareal numbers, 219 ff.
Umfänge, 110
Undecidable$_2$, 102, 215
Uniqueness theorem, 210
Unit(s), 242
  actual-infinitesimal, 222
  complex, 229
  element, 178 f.
  fractions, decomposition of, 50
  linear combination of, 229 f., 232
  negative, 36
  positive, 36
  real, 234
  segment, image of, 160, 173
Unity, 106
Universal
  arithmetic, 16
  method, 148
Upper class, 200-205
Urysohn, 173

V

Vagueness, mathematics of, 225
Varignon, 124
Vector, 226, 234
Véronese, 18, 223, 235, 242
Vicious circle, 24, 26, 76, 80
Vieta, 44
Visual
  number, 117
  space, 213, 216

W

Wallis, 20, 44 ff.
  ordering of, 47
Wave(s), 63, 154-157
  mechanics, 212
Weierstrass, 157 ff., 195 f., 233
Wittgenstein, 71, 92, 245
Wolff, 124
w-plane, 227
Writing, 51

X

x-axis, 10, 154, 156, 158

Y

y-axis, 10

Z

z-plane, 207
Zero, 103 f., 241
  dimensional, 174